SELECTIVE & OC

MATHS

REASONING

Advanced Problem Solving & Exam Strategies

ISBN: 978-1-7645316-5-8

First published 2026
Published by iSoft, Australia

Disclaimer:
This publication is independently produced and is not endorsed by, affiliated with, or approved by any selective school, Opportunity Class (OC) authority, or official testing body.

Every effort has been made to ensure the accuracy of the content in this book. However, errors or omissions may occasionally occur. Readers are encouraged to verify solutions independently and use this material as a learning support resource.

Welcome!

You're back and that's a great sign.
I'm really glad that you've decided to take the next step.

If you have worked through ***The Ultimate Guide to Selective & OC Maths Reasoning***, you have already built some strong foundations. Now it's time to take your thinking to the next level.

This book will challenge you in new ways. Some problems may take longer. Some may not make sense straight away. That's expected.

Pause. Think. Try again.
That's where real progress happens.
You don't need to rush. Give yourself time to think carefully and work through each idea.
There will be moments when things click quickly, and moments when they take longer. Both are part of learning.

Stay patient. Stay focused.
Every time you keep going, even when something feels difficult, you are becoming a stronger thinker.

This book is here to stretch your thinking and help you handle more complex problems with confidence.

Take your time. Keep working. Keep improving.

You're ready for this.

Contents

Number & Arithmetic 10

Different Remainder (÷5) 11

Remainders of Multiples (÷10) 14

Find Unknown Mass from Total 17

Linear Equation (Jar Weight Problem) 20

Proportional Reasoning (Liquid Weight) 23

Error Analysis (Inverse Operations) 26

Multiplication–Division Error Correction 29

Unit Rate with Area and Mass 32

Equivalent Weights (Ratio Comparison) 34

Minimum and Maximum Digits 37

Product of Consecutive Integers 40

Missing Amount (Multiples of 8) 43

Ordering Rational Numbers 46

Greatest 4-Digit Common Multiple 50

Least Common Multiple 53

Equivalent Ratios of Carton Sizes 57

HCF and LCM Number Pairs 60

Fractions, Decimals & Percentages 65

Fractions on a Number Line 66

Comparing Fractions and Decimals 69

Comparing Common Fractions 72

Comparing Fractions of a Whole 76

Combining Masses (unit conversion) 80

Ordering Decimals to Find Median 84

Decimal Multiplication Place Value 88

Estimating Thousandths of a Unit 91
Fractions of a Circle (spinner) 94
Finding an Original Quantity (fractions) 97
Fractions and Ratios in Groups 101

Algebra & Equations 106

Finding a Missing Divisor 107
Finding a Missing Dividend 110
Finding Missing Table Values (equations) 113
Finding Missing Totals in a Table (equations) 117
Choosing Operations in an Equation 121
Comparing Multiplicative Relationships 124
Solving Cost Equations (unit prices) 127
Finding an Unknown in an Equation 131
Brick Wall Number Patterns 134
Equivalent Fractions (fruit baskets) 137
One-Step Linear Equation (puzzle) 140
Working Out Prices from Clues 144
Finding Full Weight from Fractions 148
Working Out Meal Deal Prices 152
Finding the Starting Amount 155
System of Linear Equations (fundraising context) 158
Linear Equations (charity totals) 161
Compound Inequalities (books per box) 165
Digit Reversal (two-digit numbers) 168
Division with Remainder (word problem) 172

Geometry & Measurement 176

Triangle Area on Grid 177
Addition of Capacity (L and mL) 181
Volume of a Rectangular Prism 184
Unit Conversion (Capacity) 187

Dividing Length Measurements 190
Measuring Angles with a Protractor 193
Angle Turns (Compass Directions) 196
Turning Angles (Compass Directions) 199
Final Direction After Turns 202
Clock Hand Angle Turns 205
Finding Rectangle Length 208
Total Edge Length (Rectangular Prism) 211
Perimeter of Smaller Squares 214
Perimeter Change After Rearrangement 217
Perimeter of Enlarged Rectangle 221
Tablecloth Perimeter (one-sided overhang) 225
Area Scale Drawing (Rectangle) 228
Scale Drawing Distances (Chessboard) 231
Composite Area (Rectangle and Triangle) 234
Area of Triangle in Rectangle 237
Area Scale Factor (Squares) 240
Area Reduction in Triangles 243
Area from Equal Perimeter 246
Composite Area of Repeating Pattern 249
Composite Area (Square with Border Strips) 252
Compass Directions 255
Lines of Symmetry (Quadrilateral) 258
Reflections of a Square Tile 262
Metric Distance Conversion Errors 266
Overlapping Lengths in a Line 269
Rectangular Shadows of 3D Shapes 272
Properties of 3D Shapes 275
Painted Cube Face Counting 279
Edge Painted Cube Counting 282

Interior Cubes in a Cube 285
Hollow Cube Counting (face views) 288
Cube with Intersecting Tunnels 292
Unit Cubes in a Rectangular Block 295
Mass and Volume Scaling 298
Interior Area of a Rectangular Border 301
Packing Trapeziums in a Square 305
Liquid Volume and Mass Relationship 308
Logic & Problem Solving 312
Equivalent Equations with Number Selection 313
Minimum Button Presses (Integers) 317
Odd and Even Reasoning 321
Combinations and Sum Properties 326
Testing Statements About Totals 329
Counting Outfit Combinations 332
Distances on a Straight Line 335
Distance between points 339
Estimating Mass in Tonnes 342
Linear Combinations of Scores 345
Patterns & Sequences 348
Multiplicative Number Patterns 349
Difference Pyramid Patterns 352
Reverse Number Pattern Rules 356
Growing Number Sentence Patterns 360
Perfect Squares in Growth Patterns 363
Repeating Letter Patterns 366
Coordinate Patterns on a Grid 369
Locating Numbers in Rows and Columns 372
Common Multiples of Intervals 375
Net Movement on a Grid 379

Time, Money & Rates 384
Elapsed Time with Fractions 385
Time zone calculations 388
Time Differences Between Cities 391
Converting Times Across Time Zones 394
Converting Months to Seconds 397
Comparing Unit Price and Pack Cost 400
Cost Savings (Bulk Purchase) 403
Speed Calculation (Travel Rate) 407
Systems of Equations (Token Values) 410
Counting Coin Combinations (Change) 413
Arithmetic Sequence (reading time) 418
Rates and Resource Consumption (fish feeding) 421
Multi-Step Fraction Problem (apples) 424
Relative Time Rate (faulty clock) 427
Greatest Possible Difference (decimals) 429
Optimization of Bulk Purchasing (unit pricing) 432
Least Common Multiple (intervals) 436
Elapsed Time from Timetables (ferry trip) 439
Rates and Proportional Reasoning (rotations) 442
Combined Rates Comparison (rowing and swimming) 445
Data & Probability 449
Probability with Part-Whole Counts (stickers) 450
Weighted Total from Frequency Table 453
Average with a Missing Value 456
Probability from Ratio Relationships (stickers) 459
Conditional Probability After Removal (fruits) 462
Probability with Decimal Comparison 465
Finding a Missing Value (Line Graph) 468
Finding a Missing Frequency (Dot Plot) 471

Ordering Fractions by Size 474
Time-Series Dot Plot Analysis (ferries) 477
Categorical Data Analysis (school activities) 480
Cumulative Totals Comparison (stamps) 484
A Final Word Before You Step Forward 488

Number & Arithmetic

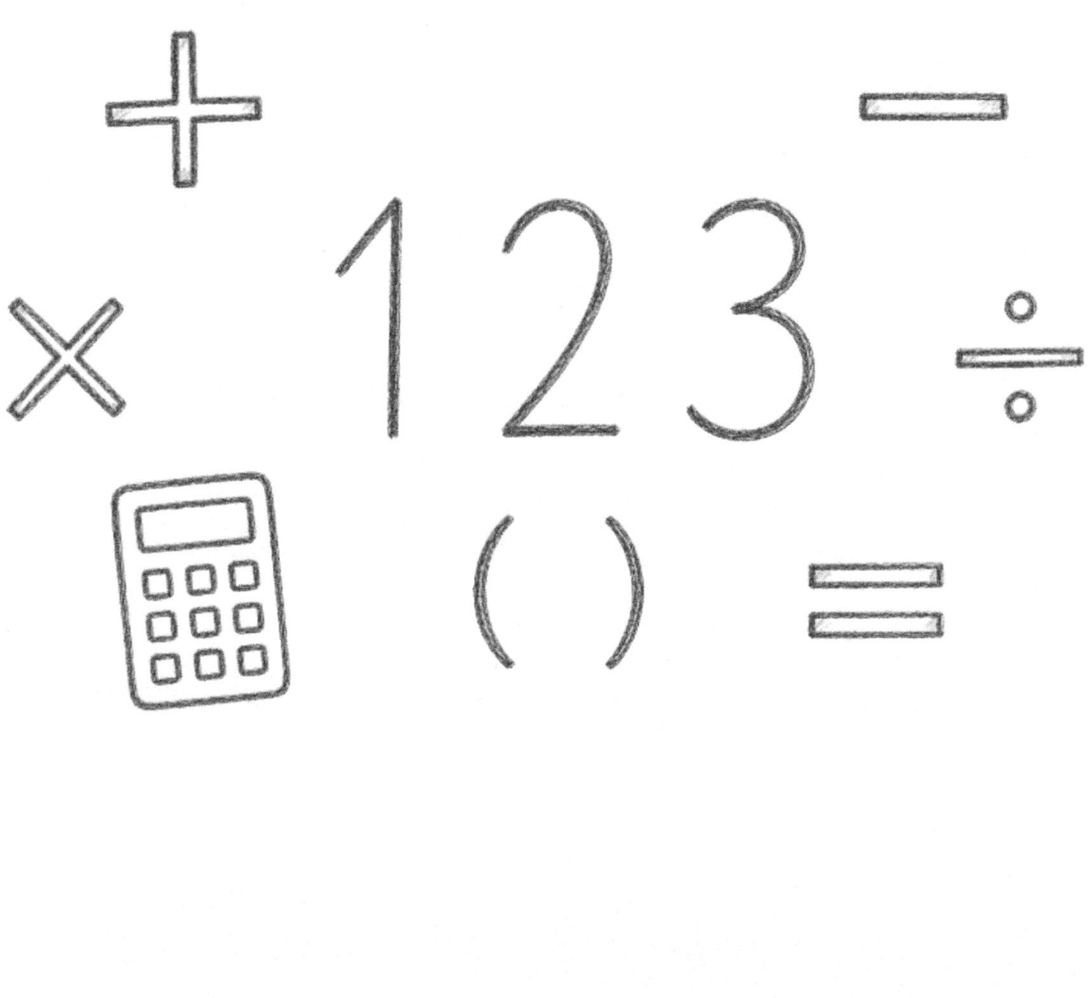

Different Remainder (÷5)

QUESTION

Five number cards show these numbers:

Four of these numbers give the same remainder when divided by 5. Which number is different from the others?

LET'S UNDERSTAND THE QUESTION TOGETHER

When we divide a number by 5, sometimes there is a leftover part. This leftover is called the **remainder**.

Here, most of the numbers behave the same when divided by 5, but one number behaves differently. Our job is to spot that odd one out.

A helpful shortcut: the remainder when **dividing by 5** depends only on the **last digit** of the number.

WHAT WE KNOW

- ✓ **Numbers given:** 31, 46, 61, 79, 101

Remainder pattern when dividing by 5

- ✓ Ends in 0 or 5 → remainder 0
- ✓ Ends in 1 or 6 → remainder 1
- ✓ Ends in 2 or 7 → remainder 2
- ✓ Ends in 3 or 8 → remainder 3
- ✓ Ends in 4 or 9 → remainder 4

WHAT WE NEED TO FIND

Which number gives a different remainder when divided by 5?

Our Plan

We will:

- ✓ Look at the last digit of each number.
- ✓ Identify its remainder when divided by 5.
- ✓ Compare all remainders to find the odd one out.

Step-by-Step Solution

Number	Remainder Explanation
31	Ends in 1 → remainder **1**
46	Ends in 6 → remainder **1**
61	Ends in 1 → remainder **1**
79	Ends in 9 → remainder **4**
101	Ends in 1 → remainder **1**

Compare remainders:

- ✓ 31 → remainder 1
- ✓ 46 → remainder 1
- ✓ 61 → remainder 1
- ✓ **79 → remainder 4**
- ✓ 101 → remainder 1

Four numbers have **remainder 1**. Only **79** has a different remainder.

Quick Check

Numbers ending in 1 or 6 usually leave remainder 1 when divided by 5. Since 79 ends in 9, getting remainder 4 makes sense.

Final Answer

79 is the number that is different from the others.

📌 Common Mistake

Students sometimes divide every number fully. Just checking the last digit is faster and safer.

💡 Exam Tip

For division by 5, always glance at the last digit first. It saves time in exams.

Remainders of Multiples (÷10)

QUESTION

A number is known to be a multiple of 4. When this number is divided by 10, what are all the possible remainders it could have?

LET'S UNDERSTAND THE QUESTION TOGETHER

A **multiple of 4** is any number that can be written in the form:

$$\boldsymbol{n = 4k}$$

where **k** is a whole number (including 0).

Examples of multiples of 4 are: **0, 4, 8, 12, 16, 20, ...**

We want to find all possible **remainders** when such numbers are **divided by 10**.

WHAT WE KNOW

- ✓ The number is a multiple of 4.
- ✓ So, we can write it as $n = 4k$.
- ✓ We are dividing the number by 10.

WHAT WE NEED TO FIND

All possible remainders when a multiple of 4 is divided by 10.

Our Plan

1. Start with $n = 4k$.
2. Calculate some multiples of 4.
3. Divide each by 10 and record the remainder.
4. Stop once the remainders start repeating.

Step-by-Step Solution

First calculate some **multiples of 4**:

$$
\begin{aligned}
4(0) &= 0 \\
4(1) &= 4 \\
4(2) &= 8 \\
4(3) &= 12 \\
4(4) &= 16 \\
4(5) &= 20
\end{aligned}
$$

Now **divide** each number **by 10**:

$0 \div 10 \rightarrow$ remainder 0
$4 \div 10 \rightarrow$ remainder 4
$8 \div 10 \rightarrow$ remainder 8
$12 \div 10 \rightarrow$ remainder 2
$16 \div 10 \rightarrow$ remainder 6
$20 \div 10 \rightarrow$ remainder 0 (pattern repeats)

We now see the repeating pattern: 0, 4, 8, 2, 6, 0, 4, 8, 2, 6, ...

The distinct possible remainders are: 0, 2, 4, 6, 8

Quick Check

Multiples of 4 are even, so their remainders mod 10 must also be even.

The remainders we found are exactly the even numbers less than 10: 0, 2, 4, 6, 8

Final Answer

When a multiple of 4 is divided by 10, the possible remainders are:

0, 2, 4, 6, 8

📌 Common Mistake

It might seem like the remainder could be any number from 0 to 9. But remember the number must be a multiple of 4.

💡 Exam Tip

When you see phrases like:

- ✓ *multiple of 4*
- ✓ *even number*
- ✓ *or multiple of 5*

Write it algebraically first, for example $n = 4k$.

Then test a few values until a pattern appears. It saves time and earns marks.

Find Unknown Mass from Total

QUESTION

Three parcels are placed on a weighing scale at the post office.

What is the mass of Parcel C?

LET'S UNDERSTAND THE QUESTION TOGETHER

All three parcels are weighed together.

We already know:

- ✓ The total mass from the scale.
- ✓ The mass of two parcels.

So, this is a **"missing part" problem**.

If we subtract the known masses from the total, we will find the mass of the third parcel.

Missing mass = Total mass − Known masses

WHAT WE KNOW

- ✓ Total mass = **3 kg**
- ✓ Parcel A = **720 g**
- ✓ Parcel B = **1.15 kg**

WHAT WE NEED TO FIND

The mass of **Parcel C**

Our Plan

To subtract properly, all masses must be in the **same unit**.

We will:

1. Convert everything to **grams**.
2. Add the known parcels.
3. Subtract from the total.

Step-by-Step Solution

Step 1: Convert everything to grams

We know:

$$1 \text{ kg} = 1000 \text{ g}$$

Total mass:

$$3 \text{ kg} = 3000 \text{ g}$$

Parcel B:

$$1.15 \text{ kg} = 1150 \text{ g}$$

Now we have:

- ✓ Total = **3000 g**
- ✓ Parcel A = **720 g**
- ✓ Parcel B = **1150 g**

Step 2: Add the known parcels

$$720 + 1150 = 1870 \text{ g}$$

Step 3: Subtract from the total

$$3000 - 1870 = 1130 \text{ g}$$

Quick Check

Let's check by adding all three:

$$720 + 1150 + 1130 = 3000\text{ g}$$

That matches the total on the scale.

Final Answer

The mass of **Parcel C is 1130 grams** (or **1.13 kg**).

📌 Common Mistake

*Forgetting to convert **kilograms to grams** before subtracting.*

✓ *Always use the **same unit** when adding or subtracting!*

💡 Exam Tip

When you see a "total weight" question, remember:

Total − Known parts = Missing part

This shortcut works every time for these types of problems!

Linear Equation (Jar Weight Problem)

QUESTION

Mika has two identical jars of honey: one full and one half-full.

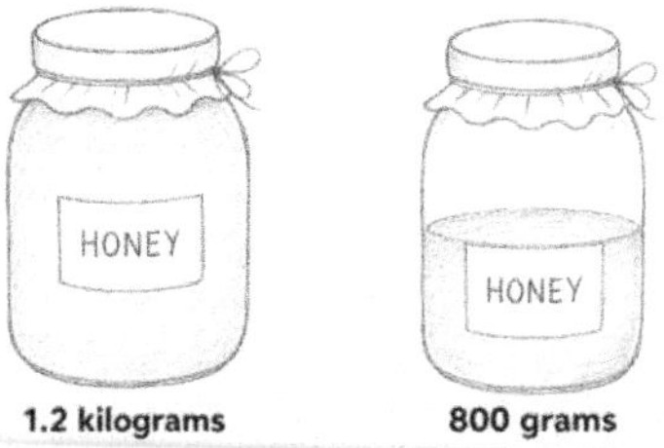

After she finishes all the honey, what does the empty jar weigh?

LET'S UNDERSTAND THE QUESTION TOGETHER

This problem is about separating the **weight of the jar** from the **weight of the honey inside**.

When the jar is full, the weight includes:

- ✓ the jar
- ✓ all the honey

When half the honey is used, the jar still weighs something, but now it contains:

- ✓ the jar
- ✓ only half the honey

If we compare the two situations carefully, we can work out the weight of the honey first. Then we can find the weight of the empty jar.

WHAT WE KNOW

- ✓ Full jar = **1.2 kilograms.**
- ✓ Half-full jar = **800 grams.**
- ✓ One jar is full, and the other contains half the honey.

WHAT WE NEED TO FIND

The weight of the **empty jar.**

Our Plan

1. Convert everything into the same unit.
2. Subtract the **half-full jar** from **the full jar**.
3. Realise that the difference represents **half the honey**.
4. Double it to find the total honey weight.
5. Subtract to find the empty jar weight.

This works because the jar's weight stays the same the whole time.

Step-by-Step Solution

Step 1: Convert units

1.2 kilograms = 1200 grams

So:

Full jar = **1200 g**
Half-full jar = **800 g**

Step 2: Subtract the weights

Full jar (with honey) - Half-full jar (with honey) = Half the honey

$1200 - 800 = 400$

That **400 g** is the weight of **half the honey**.

Step 3: Find the weight of all the honey

$400 \times 2 = 800$

So, the total honey weighs **800 grams**.

Step 4: Find the weight of the empty jar

$$1200 - 800 = 400$$

Quick Check

Full jar:

$$400 + 800 = 1200$$

Half-full jar:

$$400 + 400 = 800$$

The numbers match perfectly!

Final Answer

The empty jar weighs **400 grams.**

📌 Common Mistake

Forgetting to convert kilograms to grams before starting.
Always make sure your units match!

💡 Exam Tip

Subtract the half-full jar weight from the full jar weight. The result tells you how much honey was removed.

In other words: **Full jar** (with honey) — **Half-full jar** (with honey) = **Half the honey.**

Proportional Reasoning (Liquid Weight)

QUESTION

A metal water bottle weighs **480 g** when it is empty.

When the bottle is **one-third full of juice**, the bottle and the juice together weigh **720 g**.

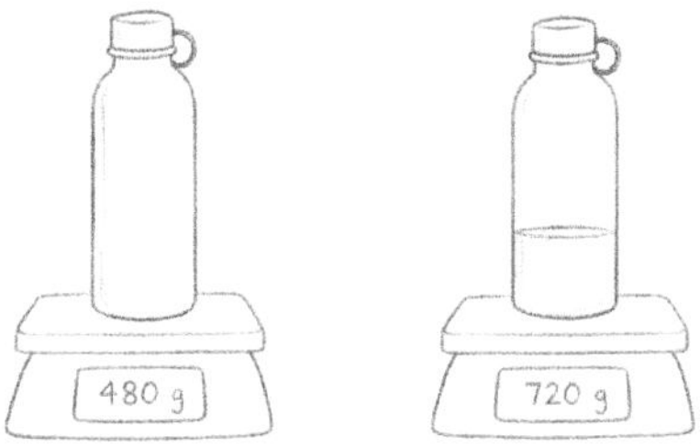

If the bottle is **two-thirds full of juice**, what is the total weight of the bottle and the juice?

LET'S UNDERSTAND THE QUESTION TOGETHER

This question is about weight.

We know how heavy the empty bottle is.
We also know how heavy it is when it is partly filled.

From that, we can figure out how much the liquid weighs and then work out the new total when the bottle is filled more.

It's like solving a puzzle step by step!

WHAT WE KNOW

- ✓ Empty bottle = **480 g**
- ✓ Bottle + one-third juice = **720 g**

WHAT WE NEED TO FIND

The total weight when the bottle is **two-thirds full**.

Our Plan

1. First, find the weight of **one-third of the juice**.
2. Then multiply to find the weight of **two-thirds of the juice**.
3. Finally, add the bottle's weight back.

We solve it in small, clear steps.

Step-by-Step Solution

Step 1: Find the weight of one-third of the juice

Bottle + one-third juice = 720 g

Empty bottle = 480 g

So, the juice alone weighs:

$720 - 480 = 240$ g

That means **one-third of the juice weighs 240 g.**

Step 2: Find the weight of two-thirds of the juice

Two-thirds is double one-third:

$240 \times 2 = 480$ g

So, two-thirds of the juice weighs **480 g.**

Step 3: Add the bottle weight

Bottle weight = 480 g

Two-thirds juice = 480 g

$480 + 480 = 960$ g

Quick Check

One-third juice = 240 g

So, the total juice when full would be:

$$240 \times 3 = 720 \text{ g}$$

Two-thirds of the juice would be:

$$720 \div 3 \times 2 = 480 \text{ g}$$

Now add the bottle weight:

$$\mathbf{480 + 480 = 960}\text{g}$$

Everything matches the earlier result

Final Answer

If the bottle is two-thirds full, it weighs **960 g** in total.

Common Mistake

Forgetting to subtract the empty bottle weight first.
Always remove the container's weight before working with the liquid.

Exam Tip

When a question involves fractions like one third or one quarter, first find the value of that fraction and then scale up carefully.

Error Analysis (Inverse Operations)

QUESTION

Liam was supposed to **subtract 85** from a number.

But by accident, he **added 85** instead.

His final answer was **310**.

What should his correct answer have been?

LET'S UNDERSTAND THE QUESTION TOGETHER

Oops! Liam did the opposite operation.

Instead of subtracting 85, he added 85.

That means his answer became **too big**.

We need to work backwards to find the correct answer.

WHAT WE KNOW

- ✓ He **should have subtracted 85.**
- ✓ He **added 85 instead.**
- ✓ His (wrong) answer was **310.**

WHAT WE NEED TO FIND

The **correct answer** after subtracting 85 properly.

Our Plan

If someone adds when they should subtract, the result is **too large by twice the number involved**.

Why twice?

Because:

- First, they went up 85 instead of down 85
- That creates a difference of $\mathbf{85 + 85 = 170}$

So, we subtract $\mathbf{2 \times 85}$ from the wrong answer.

Step-by-Step Solution

Step 1: Find double 85

$$2 \times 85 = 170$$

Step 2: Subtract from the wrong answer

$$310 - 170 = 140$$

Quick Check

Our correct answer is **140**.

If Liam made the mistake, he added 85 instead of subtracting 85.

So, his answer would increase by **170** (because $85 + 85 = 170$).

Check: $140 + 170 = 310$

This matches the wrong answer given, so **140 is correct.**

Final Answer

The correct answer should have been **140**.

Common Mistake

*Remember, the mistake made the answer **170 too large**, not just 85.*

Because instead of decreasing by 85, the value increased by 85.

*That creates a total difference of **85 + 85 = 170**.*

Exam Tip

If someone does the opposite operation, think:

Wrong answer − (2 × number involved)

Multiplication–Division Error Correction

QUESTION

Liam was supposed to **multiply a number by 30**, but he accidentally **divided** the number by 30 instead.

His incorrect answer was 8.

What answer should he have obtained if he had multiplied correctly?

LET'S UNDERSTAND THE QUESTION TOGETHER

Liam made a simple mistake.

Instead of multiplying by 30, he divided by 30.

We are told the wrong answer, which is 8.

That helps us work backwards to find the original number.

Once we know the original number, we can then work out what the correct answer should have been.

WHAT WE KNOW

- ✓ Liam divided by 30 by mistake.
- ✓ His incorrect answer was 8.
- ✓ He should have multiplied by 30.

WHAT WE NEED TO FIND

We need to find the correct answer if he had multiplied the number by 30.

Our Plan

1. Work backwards to find the original number.
2. Then multiply that number by 30.

We use inverse operations because multiplication and division undo each other.

Step-by-Step Solution

Step 1: Find the original number

He divided the number by 30 and got 8.

So:

Original number $= 8 \times 30 = 240$

The original number was **240**.

Step 2: Multiply correctly

$240 \times 30 = 7{,}200$

Quick Check

Notice this pattern:

- He **divided by 30** instead of multiplying.
- Dividing makes numbers smaller.
- Multiplying makes numbers larger.

If the incorrect answer was **8**, the correct answer must be **much larger than 8**.

Now check our answer:

Correct answer: $240 \times 30 = 7{,}200$

Mistake check: $240 \div 30 = 8$

That matches the mistake described in the question.

✓ Everything links perfectly.

Final Answer

If Liam had multiplied correctly, the answer would have been **7,200.**

📌 Common Mistake

Some students forget to find the original number first and multiply the wrong number again.

Always undo the mistake before doing the correct operation.

💡 Exam Tip

When a question says someone divided instead of multiplied, or the other way around, think about using the opposite operation first.

Inverse operations are a powerful tool in Selective-style problems!

Unit Rate with Area and Mass

QUESTION

A paper supplier packs large drawing sheets for schools.

30 sheets of this paper together cover exactly **6 square metres.**

1 square metre of this paper weighs **75 grams.**

What is the weight of one sheet of paper?

LET'S UNDERSTAND THE QUESTION TOGETHER

We are told how heavy **1 square metre** of the paper is.

We are also told that 30 sheets together cover **6 square metres.**

All the sheets are the same size and thickness.

We need to figure out how heavy just **one sheet** is.

WHAT WE KNOW

- ✓ 30 sheets cover **6 square metres.**
- ✓ 1 square metre weighs **75 grams.**
- ✓ All sheets are identical.

WHAT WE NEED TO FIND

The weight of **one sheet.**

Our Plan

1. First, find the total weight of **6 square metres**.
2. Then divide that total weight equally among the **30 sheets**.

This works because equal sheets share the total weight equally.

Step-by-Step Solution

Step 1: Find the weight of 6 square metres

$6 \times 75 = 450$

So, 6 square metres weigh **450 grams**.

Step 2: Divide by the number of sheets

The total weight of the 30 sheets is 450 grams.

Since all 30 sheets are identical, we divide the total weight equally:

$450 \div 30 = 15$

Quick Check

If one sheet weighs 15 grams:

$30 \times 15 = 450$

That matches the total weight.

Final Answer

One sheet of paper weighs 15 grams.

Common Mistake

Dividing 75 by 30 immediately.
Remember, 75 grams is for ***1 square metre****, not for all 6 square metres.*

Exam Tip

When you see area and weight together, always match the areas first before dividing.

Equivalent Weights (Ratio Comparison)

QUESTION

Lina has some identical **toy blocks** and some identical **metal balls**.

The weight of **6 toy blocks** is equal to the weight of **4 metal balls**.

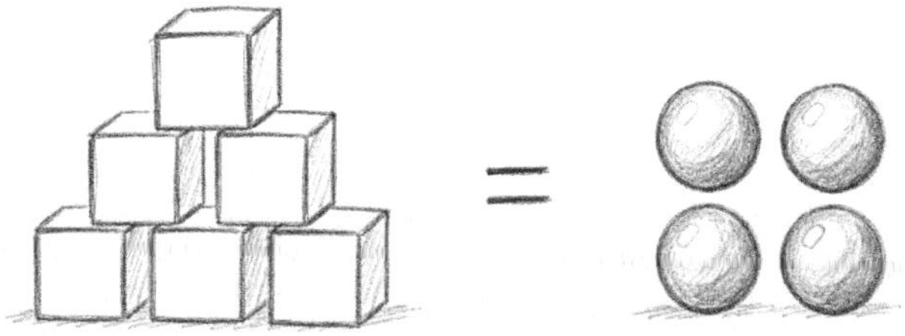

Each metal ball weighs **720 grams**.

How much does one toy block weigh?

LET'S UNDERSTAND THE QUESTION TOGETHER

This question is comparing the weight of two different objects.

We are told that a certain number of toy blocks weigh the same as a certain number of metal balls.

We already know how heavy one metal ball is.

So, the key idea is understanding how the two groups are equal in weight and then figuring out what that means for just one toy block.

WHAT WE KNOW

- ✓ 6 toy blocks = weight of 4 metal balls
- ✓ 1 metal ball weighs **720 g**

WHAT WE NEED TO FIND

The weight of **one toy block.**

Our Plan

1. First find the total weight of 4 metal balls.
2. That total equals the weight of 6 toy blocks.
3. Divide by 6 to find the weight of one toy block.

Step-by-Step Solution

Step 1: Find the weight of 4 metal balls

Each ball weighs 720 g.

$4 \times 720\ g = 2880\ g$

So, 4 metal balls weigh **2880 g**.

Step 2: This equals the weight of 6 toy blocks

So, 6 toy blocks = 2880 g

Step 3: Find the weight of one toy block

$2880\ g \div 6 = 480\ g$

So, one toy block weighs **480 g**.

Quick Check

If one toy block weighs 480 g:

$6 \times 480 = 2880$

And we already know 4 metal balls weigh 2880 g.

The weights match perfectly!

Final Answer

One toy block weighs 480 grams.

📌 Common Mistake

Some students divide 720 by 6 straight away.

***Remember**, 720 g is the weight of **one ball**, but the question compares **4 balls** to **6 blocks**.*

💡 Exam Tip

When two groups are equal in weight, always:

1. *Find the total weight of the known group first.*
2. *Then divide to find one item in the unknown group.*

Minimum and Maximum Digits

QUESTION

Mika chooses a **2-digit** number. Noah also chooses a **2-digit** number.

They **add** the two numbers together.

How many digits could their answer have?

LET'S UNDERSTAND THE QUESTION TOGETHER

This question is about adding two 2-digit numbers.

A 2-digit number can be **as small as 10** or **as large as 99**.
So the total could be fairly small or much bigger, depending on the numbers chosen.

We are not trying to find one exact sum.
We are trying to find all the possible numbers of digits in the sum.

WHAT WE KNOW

- ✓ Mika's number is a 2-digit number.
- ✓ Noah's number is a 2-digit number.
- ✓ The smallest 2-digit number is **10**.
- ✓ The largest 2-digit number is **99**.

WHAT WE NEED TO FIND

We need to find all the possible numbers of digits in the sum.

Our Plan

We will find:

- ✓ The **smallest** possible **sum.**
- ✓ The **largest** possible **sum.**

This will show the full range of answers. Then we can work out how many digits the sum could have.

Step-by-Step Solution

Step 1: Find the smallest possible sum

The smallest 2-digit number is **10.**

So, the smallest possible sum is:

$$10 + 10 = 20$$

The number **20** has **2 digits.**

Step 2: Find the largest possible sum

The largest 2-digit number is **99.**

So, the largest possible sum is:

$$99 + 99 = 198$$

The number **198** has **3 digits.**

Step 3: Work out the possible digit lengths

The sum can be as small as **20** and as large as **198.**

- ✓ Numbers from **20 to 99** have **2 digits**
- ✓ Numbers from **100 to 198** have **3 digits**

Since the possible sums **range from 20 to 198,** the answer could have **2 digits or 3 digits.**

Quick Check

Let's test one example of each:

- ✓ $10 + 10 = 20 \rightarrow$ **2 digits**
- ✓ $56 + 44 = 100 \rightarrow$ **3 digits**

So, both are possible.

Final Answer

The sum can have **2 digits or 3 digits**.

Common Mistake

Thinking the answer must always have the same number of digits. It depends on which 2-digit numbers are chosen.

Exam Tip

When a question asks for possible sizes, test the smallest case and the largest case first. This helps you find the whole range quickly.

Product of Consecutive Integers

QUESTION

Mika opened her workbook flat on the table. She saw two page numbers facing each other.

When she multiplied the two page numbers together, the answer was **2550**.

In this workbook, the **left-hand page has an even number** and the **right-hand page has the next odd number**.

What is the page number on the **left-hand page**?

LET'S UNDERSTAND THE QUESTION TOGETHER

This question is about **page numbers in an open book**.

When a book is open, the two page numbers you see are **next to each other**. That means they are **consecutive numbers**.

We are told that multiplying the two page numbers gives **2550**. We need to work out which pair of consecutive page numbers makes this product.

WHAT WE KNOW

- ✓ The book is open to two facing pages.
- ✓ Facing page numbers are **consecutive numbers**.
- ✓ The **left-hand page is even**.
- ✓ The **right-hand page is the next odd number**.
- ✓ The product of the two page numbers is **2550**.

WHAT WE NEED TO FIND

We need to find the number on the **left-hand page**.

Our Plan

We will use **estimation** first.

We will look for a square number close to **2550**. That helps us guess which number to try.

Then we will test the next consecutive number.

Step-by-Step Solution

Step 1: Think of a square number close to 2550

We need two consecutive numbers that multiply to **2550**.

A **useful** idea is to **look for a square number** close to 2550.

A student might know:

- $10 \times 10 = 100$
- $20 \times 20 = 400$
- $30 \times 30 = 900$
- $40 \times 40 = 1600$
- $50 \times 50 = 2500$

Since **2550** is very close to **2500**, the page numbers are probably close to **50**.

Step 2: Try the next consecutive number

Since **2550** is close to **2500**, and:

$$\mathbf{50 \times 50 = 2500}$$

The two page numbers are likely close to **50**.

The page numbers must be consecutive, so try:

50 and 51

Now multiply:

$$\mathbf{50 \times 51 = 2550}$$

This matches the product in the question.

Step 3: Check the page order

The question says:

- ✓ The **left-hand page is even.**
- ✓ The **right-hand page is the next odd number.**

Here:

- ✓ 50 is even
- ✓ 51 is odd

So, the page order is correct.

Step 4: State the answer

So, the left-hand page number is 50

Quick Check

$50 \times 51 = 2550$

The multiplication is correct.

Also, **50 is even and 51 is the next odd number**, so the page order makes sense.

Final Answer

The page number on the **left-hand page is 50.**

Common Mistake

Some students find the correct numbers but forget to check which one belongs on the left-hand side.

Exam Tip

When two consecutive numbers multiply to make a large number, look for a ***square number nearby*** *first.*

Missing Amount (Multiples of 8)

QUESTION

At a school fundraiser, Mika is putting cupcakes into trays.
Each tray holds **8 cupcakes**.

She wants to fill **20 trays** completely.
Mika has already baked **147 cupcakes**.

How many additional cupcakes does Mika need to bake?

LET'S UNDERSTAND THE QUESTION TOGETHER

Mika is getting cupcakes ready for the fundraiser.
She wants all the trays to be completely full.
Each tray must hold the same number of cupcakes.

The question is asking how many more cupcakes Mika needs.
If she does not, we need to find how many more cupcakes are missing.

We are not trying to find how many trays she can fill.
We are trying to find how many extra cupcakes she needs so that **all 20 trays** are full.

WHAT WE KNOW

- ✓ Each tray holds **8 cupcakes.**
- ✓ Mika wants to fill **20 trays.**
- ✓ She has already baked **147 cupcakes.**

WHAT WE NEED TO FIND

We need to find **how many more cupcakes Mika needs.**

Our Plan

First, we will find the **total number of cupcakes** needed to fill all 20 trays.
Then, we will compare that with the **147 cupcakes** Mika already has.
The difference will tell us how many more cupcakes she needs.

Step-by-Step Solution

Step 1: Find the total number of cupcakes needed

Each tray holds 8 cupcakes, and there are 20 trays.

$$20 \times 8 = 160$$

So, Mika needs **160 cupcakes** altogether.

Step 2: Find how many more cupcakes she needs

Mika needs **160 cupcakes** altogether.

She already has **147 cupcakes**, so we subtract:

$$\mathbf{160 - 147 = 13} \textbf{ cupcakes}$$

So, Mika needs **13 more cupcakes.**

Quick Check

Let's check our answer.

Mika has **147** cupcakes and needs **13** more.

$$147 + 13 = 160$$

And:

$$160 \div 8 = 20$$

So, 160 cupcakes are exactly enough to fill 20 trays.
Our answer makes sense.

Final Answer

Mika needs **13 more cupcakes**.

📌 Common Mistake

A common mistake is subtracting before finding the total number needed. Always work out the full amount needed first.

💡 Exam Tip

When you see "how many more," think:
total needed compared with the amount already there.

Ordering Rational Numbers

QUESTION

A student wrote these **4 numbers** on a card:

$$\frac{11}{5} \quad 2.08 \quad 2\frac{3}{20} \quad 2.2$$

Put the numbers **in order** from **smallest to largest**.

LET'S UNDERSTAND THE QUESTION TOGETHER

This question is about **ordering numbers**, even when they are written in different ways.

- ✓ Some are written as fractions.
- ✓ Some are decimals.
- ✓ One is a mixed number.

To compare them fairly, we can change them all into **decimals** first.
Then we can see which is smallest and which is biggest.

WHAT WE KNOW

- ✓ The four numbers are:
  - $\frac{11}{5}$
  - 2.08
  - $2\frac{3}{20}$
  - 2.2
- ✓ We need to arrange them from **smallest to largest**.

WHAT WE NEED TO FIND

We need the correct order of the numbers from **smallest to biggest**.

Our Plan

We will **convert** each number into a **decimal**.

This works well because decimals are easy to compare using place value:

- ✓ Ones
- ✓ Tenths
- ✓ Hundredths

After that, we will:

1. order the decimal values
2. write the answer again using the **original forms**

Step-by-Step Solution

Step 1: Convert each number to a decimal: $\frac{11}{5}$

$11 \div 5 = 2.2$

So,

$\frac{11}{5} = \mathbf{2.2}$

This is already a decimal: $\mathbf{2.08}$

First convert the fraction part: $2\frac{\mathbf{3}}{\mathbf{20}}$

$\frac{3}{20} = 0.15$

Now add it to 2:

$2 + 0.15 = 2.15$

So,

$$2\frac{3}{20} = \mathbf{2.15}$$

This is already a decimal: $\mathbf{2.2}$

To compare more easily, we can write it as: 2.20

Step 2: Write the decimal sequence

Now the numbers are:

- ✓ $\frac{11}{5} = 2.20$
- ✓ $2.08 = 2.08$
- ✓ $2\frac{3}{20} = 2.15$
- ✓ $2.2 = 2.20$

So, the decimal sequence from smallest to largest is:

2.08, 2.15, 2.20, 2.20

Step 3: Write the original numbers in order

Now match those decimals back to the original numbers:

- 2.08 stays as 2.08
- $2.15 = 2\frac{3}{20}$
- $2.20 = \frac{11}{5}$
- $2.20 = 2.2$

So, the original numbers in order are:

$\mathbf{2.08, 2\frac{3}{20}, \frac{11}{5}, 2.2}$

Because the last two numbers are equal, they could also be written the other way around.

Quick Check

Let's check the sizes:

- ✓ 2.08 is just a bit more than 2
- ✓ 2.15 is bigger than 2.08
- ✓ 2.20 is bigger than 2.15

That makes sense.

Also,

$$\frac{11}{5} = 2.2$$

So those last two numbers are equal.

Final Answer

The decimal sequence is:

$$2.08, 2.15, 2.20, 2.20$$

So, the original numbers from smallest to largest are:

$$2.08, 2\frac{3}{20}, \frac{11}{5}, 2.2$$

Common Mistake

Forgetting that 2.2 *is the same as* 2.20*. Adding the zero helps you compare it properly with numbers like* 2.15 *and* 2.08*.*

Exam Tip

When numbers are written in different forms, convert them all into the same form first. Decimals are usually the easiest for ordering.

Greatest 4-Digit Common Multiple

QUESTION

Sophie is looking for the greatest four-digit number that is a **multiple (can be divided exactly)** of both 3 and 10.

She then adds the digits of that number together.

What total does she get?

LET'S UNDERSTAND THE QUESTION TOGETHER

We need to find a number that follows two rules at the same time.

It must:

- ✓ Be a four-digit number
- ✓ Be a multiple (can be divided exactly) of 3
- ✓ Be a multiple (can be divided exactly) of 10

Then we add its digits.

WHAT WE KNOW

- ✓ We want the **greatest four-digit number.**
- ✓ It must be a **multiple (can be divided exactly)** of **3.**
- ✓ It must also be a **multiple (can be divided exactly)** of **10.**
- ✓ Then we add its digits.

WHAT WE NEED TO FIND

We need to find the **sum of the digits** of the greatest four-digit number that is a multiple (can be divided exactly) of both 3 and 10.

Our Plan

A number that is a **multiple of 10** must end in **0**.

A number that is a multiple of 3 has **digits that add to a multiple of 3**.

So, we will:

1. Find the greatest four-digit number ending in 0.
2. Check whether it is also a multiple of 3.
3. Add its digits.

Step-by-Step Solution

Step 1: Find the greatest four-digit number.

The greatest four-digit number is: **9999**

Step 2: Find the greatest four-digit number that is a multiple of 10.

A number that is a multiple of 10 must end in **0**.

So, the greatest four-digit number ending in 0 is: **9990**

Step 3: Check if 9990 is also a multiple of 3.

Add the digits:

$$9 + 9 + 9 + 0 = 27$$

Since **27** is a multiple of 3, **9990** is also a multiple of 3.

Step 4: Add the digits of the number

The number is **9990**.

$$9 + 9 + 9 + 0 = 27$$

So, the digit sum is **27**.

Quick Check

- ✓ 9990 is a four-digit number.
- ✓ It ends in 0, so it is a multiple of 10.
- ✓ Its digits add to 27, so it is a multiple of 3.

So, our answer makes sense.

Final Answer

The total of the digits is **27**.

Common Mistake

Choosing a number that ends in 0 but forgetting to check if it is also a multiple of 3.

Exam Tip

Use divisibility rules to save time:

- ✓ *Ends in 0 → multiple of 10*
- ✓ *Digit sum is a multiple of 3 → multiple of 3*

Least Common Multiple

QUESTION

A school has some reward tokens to pack for a maths fair.

The tokens can be placed into:

- ✓ 2 equal bags
- ✓ 4 equal bags
- ✓ 6 equal bags
- ✓ 9 equal bags
- ✓ 15 equal bags

Each time, all the tokens fit exactly, with no tokens left over.

What is the smallest number of reward tokens the school could have?

LET'S UNDERSTAND THE QUESTION TOGETHER

This question is asking for one number that works in several different ways.

The total number of tokens must split evenly into 2 bags, 4 bags, 6 bags, 9 bags, and 15 bags.

We want the **smallest** number that can do all of that with **no tokens left over**.

WHAT WE KNOW

- ✓ The tokens can be shared into **2 equal bags.**
- ✓ The tokens can be shared into **4 equal bags.**
- ✓ The tokens can be shared into **6 equal bags.**
- ✓ The tokens can be shared into **9 equal bags.**
- ✓ The tokens can be shared into **15 equal bags.**
- ✓ There are **no tokens left over.**

WHAT WE NEED TO FIND

We need to find the **smallest number of tokens** that is divisible by: 2, 4, 6, 9, 15

Our Plan

We will use this rule:

1. Cross out covered numbers.
2. If one number covers the rest, use it.
3. If not, keep the biggest pieces and multiply.

Here, **covered** means a bigger number already includes what a smaller number needs.

For example, **4 covers 2** because any number that can be divided evenly by 4 can also be divided evenly by 2.

We will also use **prime factors**.

Prime factors are the prime numbers that multiply together to make a number.

For example:

- $4 = 2 \times 2$
- $6 = 2 \times 3$
- $9 = 3 \times 3$
- $15 = 3 \times 5$

After finding the prime factors of each leftover number, we look across all the lines. Then we keep:

- the most **2s**
- the most **3s**
- the most **5s**

Then we multiply them together.

Step-by-Step Solution

Step 1: Write the numbers

The total number of tokens must be divisible by **2, 4, 6, 9, 15**

Step 2: Remove numbers that are factors of other numbers

Check if any number is a factor of another number in the list.

2 is a factor of 4, because: $4 \div 2 = 2$

So, any number divisible by **4** will also be divisible by **2**.

That means we can remove **2**.

We are now left with **4, 6, 9, 15**

Step 3: Check if one number is a multiple of all the others

Now look at the remaining numbers: **4, 6, 9, 15**

We check if any one of these numbers can be divided evenly by all the others.

- ✓ Does **4** divide by 6, 9, and 15? No
- ✓ Does **6** divide by 4, 9, and 15? No
- ✓ Does **9** divide by 4, 6, and 15? No
- ✓ Does **15** divide by 4, 6, and 9? No

So, no single number is a multiple of all the others. This means we need to move to the next step and use prime factors.

Step 4: Find the most 2s, 3s and 5s

First, write the prime factors of each leftover number:

- ✓ $4 = 2 \times 2$
- ✓ $6 = 2 \times 3$
- ✓ $9 = 3 \times 3$
- ✓ $15 = 3 \times 5$

Now look across all the lines:

- ✓ The most **2s** are in **4**, so keep 2×2
- ✓ The most **3s** are in **9**, so keep 3×3
- ✓ The **5** is in **15**, so keep 5

Step 5: Now multiply:

$$2 \times 2 \times 3 \times 3 \times 5$$

$$= 4 \times 9 \times 5$$

$$= 36 \times 5$$

$$= 180$$

So, the smallest number of reward tokens is: 180

Quick Check

Let's check:

✓ $180 \div 2 = 90$

✓ $180 \div 4 = 45$

✓ $180 \div 6 = 30$

✓ $180 \div 9 = 20$

✓ $180 \div 15 = 12$

Each answer is a whole number, so 180 works perfectly.

Final Answer

The smallest number of reward tokens the school could have is **180**.

Common Mistake

Multiplying all the numbers together. That gives an answer much bigger than needed.

Exam Tip

Before multiplying, always check whether some numbers can be crossed out first.

Equivalent Ratios of Carton Sizes

QUESTION

A big carton of juice contains **60 juice boxes**.

It takes **3 big cartons** to hold the same number of juice boxes as **4 medium cartons**.
It also takes **2 medium cartons** to hold the same number as **5 small cartons**.

How many juice boxes are in **1 small carton**?

LET'S UNDERSTAND THE QUESTION TOGETHER

This question is about comparing **different carton sizes**.

We know how many juice boxes are in a big carton.
Then we are told that some groups of big, medium, and small cartons hold the **same total number of juice boxes**.

So, the cartons are linked like this:

- ✓ big cartons match medium cartons
- ✓ medium cartons match small cartons

We use these links to work step by step from **big** to **small**.

WHAT WE KNOW

- ✓ 1 big carton contains **60 juice boxes.**
- ✓ **3 big cartons = 4 medium cartons.**
- ✓ **2 medium cartons = 5 small cartons.**

WHAT WE NEED TO FIND

We need to find how many juice boxes are in **1 small carton**.

Our Plan

We will:

1. Find the number of juice boxes in **3 big cartons.**
2. Use that to find the number in **1 medium carton.**
3. Find the number in **2 medium cartons.**
4. Use that to find the number in **1 small carton.**

This works because the question tells us the groups hold the same total number of juice boxes.

Step-by-Step Solution

Step 1: Find how many juice boxes are in 3 big cartons

Each big carton has 60 juice boxes.

$3 \times 60 = 180$

So, **3 big cartons contain 180 juice boxes.**

Step 2: Find how many juice boxes are in 1 medium carton

We know:

3 big cartons = 4 medium cartons

So, 4 medium cartons contain 180 juice boxes.

$180 \div 4 = 45$

So, **1 medium carton contains 45 juice boxes.**

Step 3: Find how many juice boxes are in 2 medium cartons

$2 \times 45 = 90$

So, **2 medium cartons contain 90 juice boxes.**

Step 4: Find how many juice boxes are in 1 small carton

We know:

$$2 \text{ medium cartons} = 5 \text{ small cartons}$$

So, 5 small cartons contain 90 juice boxes.

$$90 \div 5 = 18$$

So, **1 small carton contains 18 juice boxes**.

Quick Check

✓ 5 small cartons: $\mathbf{5} \times \mathbf{18} = \mathbf{90}$

✓ 2 medium cartons: $\mathbf{2} \times \mathbf{45} = \mathbf{90}$

The totals match, so the answer makes sense.

Final Answer

1 small carton contains **18** juice boxes.

Common Mistake

A common mistake is trying to divide too early.

Remember that the groups are equal:

✓ ***3 big cartons = 4 medium cartons***

✓ ***2 medium cartons = 5 small cartons***

Always find the total number of juice boxes in the matching group first, then divide to find one carton.

Exam Tip

Follow the size chain in order: ***big → medium → small.*** *This makes the working much easier to organise.*

HCF and LCM Number Pairs

QUESTION

Two numbers have a highest common factor of **3** and a lowest common multiple of **18**.

Find **all possible pairs** of numbers that satisfy these conditions.

LET'S UNDERSTAND THE QUESTION TOGETHER

This question gives us two important clues.

1. Their highest common factor is 3

The **highest common factor** is the biggest number that divides both numbers exactly.

For example, look at **3** and **18**.

Factors of **3**: 1,3

Factors of **18**: 1,2,3,6,9,18

The common factors are: 1,3

So, the highest common factor is: 3

This means both numbers in each pair must be in the **3 times table**.

2. Their lowest common multiple is 18

The **lowest common multiple** is the first number that appears in both times tables.

Multiples of **3**: 3,6,9,12,15,18, ...

Multiples of **18**: 18,36,54, ...

The first common multiple is: 18

So, the lowest common multiple is: 18

So, the question is asking us to find **every pair of numbers** that:

✓ Has HCF **3**

✓ Has LCM **18**

WHAT WE KNOW

- ✓ HCF = **3**
- ✓ LCM = **18**
- ✓ We need to find **all possible pairs**

WHAT WE NEED TO FIND

We need to find all pairs of numbers whose:

- ✓ Highest common factor is **3**
- ✓ Lowest common multiple is **18**

Our Plan

We will:

1. Divide the LCM by the HCF.
2. Find factor pairs of the result.
3. Multiply each number in the pair by 3.

Step-by-Step Solution

Step 1: Divide the LCM by the HCF

$18 \div 3 = 6$

So now we need to work with **6**.

Step 2: Find the factor pairs that make 6

We need two whole numbers that multiply to make **6**.

The factor pairs are:

1×6
2×3

So, the factor pairs of **6** are:

(1, 6) and **(2, 3)**

Now check that the two numbers in each pair do not share a common factor bigger than **1**.

✓ **1 and 6** have HCF **1**, so keep this pair.
✓ **2 and 3** have HCF **1**, so keep this pair.

So, both factor pairs can be used.

Step 3: Multiply each factor pair by 3

Since the HCF is **3**, multiply each number in each factor pair by **3**.

For 1 × 6:

$$1 \times 3 = 3$$

$$6 \times 3 = 18$$

So, one pair is: (3, 18)

For 2 × 3:

$$2 \times 3 = 6$$

$$3 \times 3 = 9$$

So, the other pair is: (6, 9)

Step 4: Write all possible pairs

The two possible pairs are: (3,18) and (6,9)

Quick Check

Check (3, 18)

HCF

Factors of 3: **1** , **3**

Factors of 18: **1**, 2, **3**, 6, 9, 18

The **common** factors are: **1**, **3**

So, the **HCF** is **3**.

LCM

Multiples of 3: 3 ,6, 9, 12, 15, **18**, ...

Multiples of 18: **18** ,36, 54, ...

The first common multiple is **18**.

Check (6, 9)

HCF

Factors of 6: **1**, 2, **3**, 6

Factors of 9: **1**, **3**, 9

The **common** factors are: **1**, **3**

So, the **HCF** is **3**.

LCM

Multiples of 6: 6, 12, **18**, 24, ...

Multiples of 9: 9, **18**, 27, ...

The first common multiple is **18**.

So, both pairs are correct.

Final Answer

The possible pairs are **(3, 18)** and **(6, 9)**.

Common Mistake

Some students find the factor pairs that make 6 but forget to multiply both numbers by 3.

Exam Tip

When HCF and LCM are given, dividing the LCM by the HCF is a very helpful first step.

Fractions, Decimals & Percentages

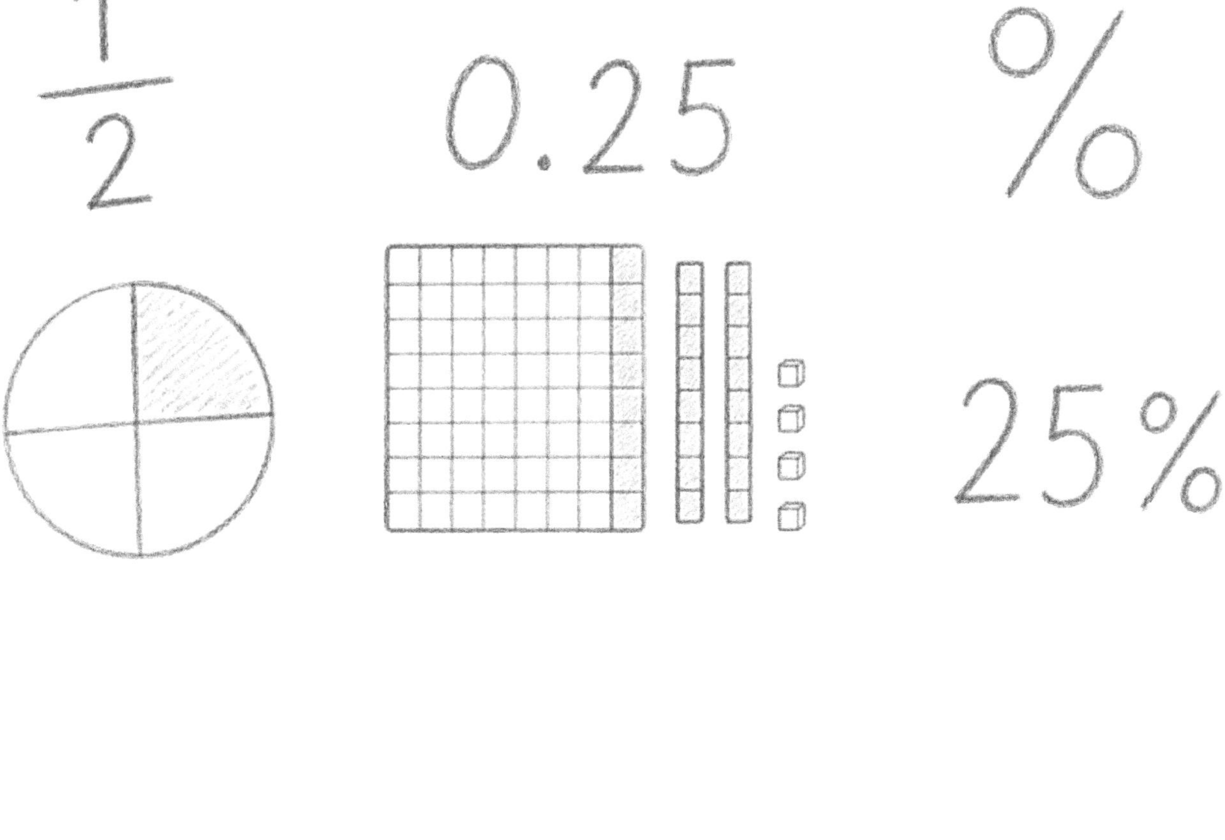

Fractions on a Number Line

QUESTION

On a number line, the marks are equally spaced.

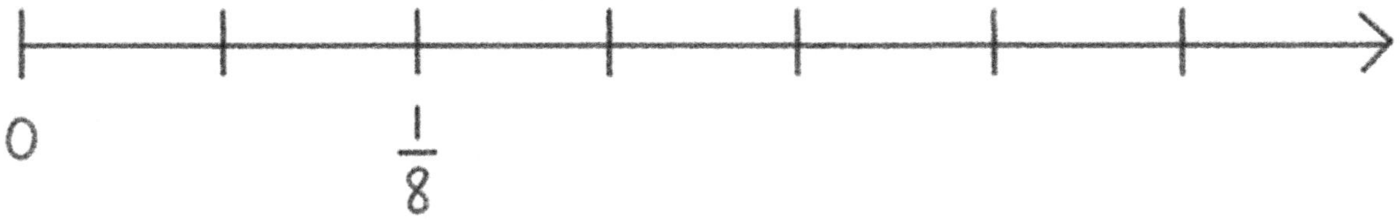

What fraction is at the **fourth mark**?

LET'S UNDERSTAND THE QUESTION TOGETHER

This question is about a **number line with equal spacing**.

That means each jump from one mark to the next is the same size.

We are given the value at one mark, and we need to work out the value at another mark.

So, this is really a question about **finding missing fractions using equal intervals**.

WHAT WE KNOW

- ✓ The marks are **equally spaced**.
- ✓ The **first mark** is **0**.
- ✓ The **third mark** is $\frac{1}{8}$.

WHAT WE NEED TO FIND

We need to find the fraction at the **fourth mark**.

We count the marks starting from 0:

- ✓ 1st mark = 0
- ✓ 2nd mark = next tick
- ✓ 3rd mark = $\frac{1}{8}$
- ✓ 4th mark = the next tick after $\frac{1}{8}$

Our Plan

We will first find the value of **one interval**.

Then we will add one more interval to the value at the third mark.

Step-by-Step Solution

Step 1: Find how many equal intervals there are from the first mark to the third mark.

From the first mark to the third mark, there are **2 equal intervals**.

So: 2 intervals $= \frac{1}{8}$

Step 2: Find the value of 1 interval.

$\frac{1}{8} \div 2 = \frac{1}{16}$

So, each step is: $\frac{1}{16}$

Step 3: Add one more interval to the third mark.

The fourth mark is one step after the third mark, so:

$$\frac{1}{8}+\frac{1}{16}$$

$$=\frac{2}{16}+\frac{1}{16}$$

$$=\frac{3}{16}$$

Step 4: Write the answer.

So, the fraction at the fourth mark is: $\frac{3}{16}$

Quick Check

The fourth mark is one step after the third mark.

So, if we subtract one interval, we should get the third mark:

$$\frac{3}{16} - \frac{1}{16} = \frac{2}{16} = \frac{1}{8}$$

This matches the given value, so the answer makes sense.

Final Answer

The fraction at the fourth mark is $\frac{3}{16}$.

📌 Common Mistake

*A common mistake is forgetting that there are **2 intervals** from the first mark to the third mark.*

💡 Exam Tip

*When marks are equally spaced, always find the value of **one interval** first.*

Comparing Fractions and Decimals

QUESTION

A recipe needs $\frac{5}{8}$ of a cup of milk.

Which of these amounts is **nearest** to $\frac{5}{8}$ of a cup?

1. 0.5 cups
2. 0.6 cups
3. 0.7 cups
4. 0.9 cups

LET'S UNDERSTAND THE QUESTION TOGETHER

This question shows us a **fraction** and some **decimal numbers**.

It asks us to find which decimal is **closest** to the fraction.

We are just comparing values to find the nearest one.

WHAT WE KNOW

- ✓ The fraction is $\frac{5}{8}$
- ✓ The answer choices are:
    1. **0.5**
    2. **0.6**
    3. **0.7**
    4. **0.9**

WHAT WE NEED TO FIND

We need to find **which decimal is nearest to** $\frac{5}{8}$.

Our Plan

First, we will change the **fraction into a decimal**.

Then we will **compare** the answer choices and see which one is closest.

After that, we can check our answer carefully.

Step-by-Step Solution

Step 1: Change the fraction into a decimal.

$\frac{5}{8} = 5 \div 8 = 0.625$

Step 2: Write the answer choices in order.

0.5

0.6

0.7

0.9

Step 3: We can see that **0.625 is between 0.6 and 0.7**.

Now compare how far it is from each:

- ✓ Distance to $0.6 = 0.625 - 0.6 = 0.025$
- ✓ Distance to $0.7 = 0.7 - 0.625 = 0.075$

Since **0.025 is smaller, 0.6 is close**

Quick Check

Now let us check properly.

Distance from **0.6**:

$0.625 - 0.6 = 0.025$

Distance from **0.7**:

$0.7 - 0.625 = 0.075$

Since **0.025** is smaller than **0.075, 0.6** is closer.

We do not need to check **0.5** and **0.9** because they are clearly farther away.

Final Answer

The number nearest to $\frac{5}{8}$ is **0.6**.

📌 Common Mistake

*Some students choose **0.7** because it also looks close, but **0.6** is actually nearer.*

💡 Exam Tip

If two answer choices look close, do a quick subtraction check to make sure you pick the nearest one.

Comparing Common Fractions

QUESTION

Which of these statements is correct?

A $\frac{2}{5}+\frac{1}{5}$ **is less than** $\frac{4}{5}$

B $1-\frac{1}{4}$ is **equal to** $\frac{3}{4}$

C $\frac{1}{8}$ is **greater than** $\frac{1}{6}$

LET'S UNDERSTAND THE QUESTION TOGETHER

This question is about **working with fractions** and checking whether each statement is true or false.

We need to look at each statement carefully.

- ✓ One statement uses **addition**
- ✓ One uses **subtraction**
- ✓ One asks us to **compare two fractions**

Our job is to test each one and see which statements are correct.

WHAT WE KNOW

Statement A: $\frac{2}{5}+\frac{1}{5}<\frac{4}{5}$

Statement B: $1-\frac{1}{4}=\frac{3}{4}$

Statement C: $\frac{1}{8}>\frac{1}{6}$

WHAT WE NEED TO FIND

We need to find **which statement or statements are correct**.

Our Plan

We will check each statement one by one.

- ✓ For **A**, we will add the fractions.
- ✓ For **B**, we will subtract from 1.
- ✓ For **C**, we will compare the fractions using **cross multiplication**.

This works because once we find the actual values, we can clearly tell if each statement is true.

Step-by-Step Solution

Step 1: Check Statement A

$$\frac{2}{5} + \frac{1}{5} = \frac{3}{5}$$

Now compare $\frac{3}{5}$ and $\frac{4}{5}$.

They have the same denominator, so we compare the numerators:

$$3 < 4$$

So,

$$\frac{3}{5} < \frac{4}{5}$$

This means **Statement A is correct**.

Step 2: Chcck Statement B

$$1 - \frac{1}{4}$$

Write 1 as quarters:

$$1 = \frac{4}{4}$$

Now subtract:

$$\frac{4}{4} - \frac{1}{4} = \frac{3}{4}$$

So, the statement is true.

This means **Statement B is correct.**

Step 3: Check Statement C

We need to compare:

$$\frac{1}{0} \text{ and } \frac{1}{6}$$

To compare them, we can use **cross multiplication.**

Multiply the numerator of the first fraction by the denominator of the second:

$$1 \times 6 = 6$$

Multiply the numerator of the second fraction by the denominator of the first:

$$1 \times 8 = 8$$

Now compare the answers:

$$6 < 8$$

So,

$$\frac{1}{8} < \frac{1}{6}$$

But the statement says:

$$\frac{1}{8} > \frac{1}{6}$$

That is not true.

So, **Statement C is incorrect.**

Quick Check

✓ $\frac{2}{5} + \frac{1}{5} = \frac{3}{5}$, and $\frac{3}{5} < \frac{4}{5}$

✓ $1 - \frac{1}{4} = \frac{3}{4}$

Cross multiplication shows $\frac{1}{8} < \frac{1}{6}$, so Statement C is **false**

Everything makes sense.

Final Answer

Statements A and B are correct.

Common Mistake

When comparing fractions like $\frac{1}{8}$ and $\frac{1}{6}$, some students think $\frac{1}{8}$ is bigger because 8 is bigger than 6. But with fractions that have 1 on top, a bigger denominator means smaller parts.

Exam Tip

Cross multiplication is a great way to compare fractions quickly, especially when the denominators are different.

Comparing Fractions of a Whole

QUESTION

Three friends, Maya, Leo and Zara, are sharing a fruit tart.

The tart is cut into **8 equal slices**.

- ✓ Maya eats **2/8** of the tart.
- ✓ Leo eats **3/8** of the tart.
- ✓ Zara eats **1/2** of the tart.

Which of these statements is correct?

1. They eat exactly one whole tart.
2. Maya eats the smallest amount.
3. Leo eats twice as much as Maya.

LET'S UNDERSTAND THE QUESTION TOGETHER

We are trying to **understand the information first**, not solve it straight away.

Three friends are sharing one fruit tart.
Each person eats a fraction of the same tart.

The question is asking us to compare those fractions carefully.
We also need to check whether the total amount makes one whole tart.

A good first step is to make the fractions easier to compare.

WHAT WE KNOW

- ✓ Maya eats **2/8** of the tart.
- ✓ Leo eats **3/8** of the tart.
- ✓ Zara eats **1/2** of the tart.
- ✓ All fractions are parts of the same whole tart.
- ✓ We need to test each statement.

WHAT WE NEED TO FIND

We need to work out **which statement is correct.**

Our Plan

We will change the fractions so they are all written in **eighths**.

This will help us compare the amounts easily because each fraction will be talking about the same sized parts of the tart.

Then we can check:

- ✓ Who eats the least.
- ✓ Whether they eat one whole tart.
- ✓ Whether one person eats twice as much as another.

Step-by-Step Solution

Step 1: Change $\frac{1}{2}$ into eighths

Zara eats: $\frac{1}{2}$

We want to write this fraction in **eighths**. To change the denominator from **2** to **8**, we multiply by **4**.

So, we multiply both the numerator and denominator by **4**:

$$\frac{1}{2} = \frac{1 \times 4}{2 \times 4} = \frac{4}{8}$$

Multiply the numerator (top) and denominator (bottom) by 4 to keep the fraction equal.

Now the fractions are:

- ✓ Maya = **2/8**
- ✓ Leo = **3/8**
- ✓ Zara = **4/8**

Step 2: Check Statement 1

They eat the whole tart.

Add the fractions:

$$\frac{2}{8}+\frac{3}{8}+\frac{4}{8}=\frac{9}{8}$$

The **numerator** (top number) is **bigger** than the **denominator** (bottom number), so the fraction is more than 1 whole.

$$\frac{9}{8}=1\frac{1}{8}$$

This means they eat 1 whole tart and 1 extra slice out of 8.

Since $\frac{9}{8}>1$, they eat more than one whole tart, so the statement is false.

.

Step 3: Check Statement 2

Maya eats the smallest amount.

Compare the fractions:

- ✓ Maya $=\frac{2}{8}$
- ✓ Leo $=\frac{3}{8}$
- ✓ Zara $=\frac{4}{8}$

Since all the fractions are in eighths, we **only** need to **compare the numerators.**

$$2<3<4$$

So:

$$\frac{2}{8}<\frac{3}{8}<\frac{4}{8}$$

Maya's fraction is the smallest.

Therefore, **Statement 2 is true**.

Step 4: Check Statement 3

Leo eats twice as much as Maya.

Maya eats: $\frac{2}{8}$

Twice as much means multiply by 2:

$$2 \times \frac{2}{8} = \frac{4}{8}$$

So twice as much as Maya would be: $\frac{4}{8}$

But Leo eats: $\frac{3}{8}$

Since: $\frac{3}{8} \neq \frac{4}{8}$

Leo does **not** eat twice as much as Maya.

Therefore, **Statement 3 is false**.

Quick Check

Let's quickly check:

- ✓ Total eaten is $\frac{9}{8}$, so it is not exactly one whole.
- ✓ $\frac{2}{8}$ is smaller than $\frac{3}{8}$ and $\frac{4}{8}$.
- ✓ Twice of $\frac{2}{8}$ is $\frac{4}{8}$, not $\frac{3}{8}$.

Final Answer

Only Statement 2 is correct. (Maya eats the smallest amount)

Common Mistake

*Forgetting to change **1/2** into **4/8** before comparing.*

Exam Tip

When fractions have different denominators, change them to the same denominator first. It makes comparing much easier.

Combining Masses (unit conversion)

QUESTION

Luca has four packets of rice with these masses:

He wants to combine **two or more** packets.

Which of these total masses can he make?

- ✓ **950 grams**
- ✓ **1.1 kilograms**
- ✓ **1 400 grams**

LET'S UNDERSTAND THE QUESTION TOGETHER

We need to find which total masses Luca can make by combining **two or more packets** of rice.

The packet masses are: $600g,\ 350g,\ 200g,\ 450g$

The target masses are: $950g,\ 1.1kg,\ 1400g$

WHAT WE KNOW

- ✓ Luca has 4 packets of rice.
- ✓ The masses are $600g,\ 350g,\ 200g,\ 450g$.
- ✓ He must combine **two or more packets.**
- ✓ We need to check which target masses can be made exactly.

WHAT WE NEED TO FIND

We need to find **which of the target masses** Luca can make exactly by combining **two or more** packets.

Our Plan

1. First, convert **everything into grams**.
2. Then find the **total mass** of all four packets.
3. After that, use the shortcut method: Instead of adding many combinations, check **what mass must be left out**.

Step-by-Step Solution

Step 1: Convert everything to grams

We know: $1kg = 1000g$

So: $1.1kg = 1100g$

Now the target masses are: $950g,\ 1100g,\ 1400g$

Step 2: Find the total mass of all packets

$600 + 350 + 200 + 450 = 1600$

So, all four packets together weigh: $1600g$

Step 3: Can Luca make $950g$?

Work backwards: $1600 - 950 = 650$

So, Luca needs to leave out $650g$.

Can he leave out $650g$?

$200 + 450 = 650$

Yes.

So, the packets he uses are: $600 + 350 = 950$

Luca **can make** $950g$.

Step 4: Can Luca make $1100g$?

Work backwards:

$$1600 - 1100 = 500$$

So, Luca needs to leave out $500g$.

Can he leave out $500g$?

No packet or combination of packets makes $500g$.

So, Luca cannot make $1100g$.

Since: $1.1kg = 1100g$

Luca **cannot** make $1.1kg$.

Step 5: Can Luca make $1400g$?

Work backwards:

$$1600 - 1400 = 200$$

So, Luca needs to leave out $200g$.

Can he leave out $200g$?

Yes, he can leave out the $200g$packet.

So, the packets he uses are:

$$600 + 350 + 450 = 1400$$

Luca **can make** $1400g$.

Quick Check

Let's check our answers:

- $600 + 350 =$ **950** ✓
- No combination equals **1100** ✗
- $600 + 350 + 450 =$ **1400** ✓

Our answers make sense.

Final Answer

The total masses Luca can make are **950 grams** and **1 400 grams**, but **not 1.1 kilograms**.

📌 Common Mistake

*Forgetting to convert **1.1 kg** into **1100 g** before starting.*

💡 Exam Tip

*When there are only 4 numbers, list all the **2-packet totals** first, then all the **3-packet totals**. This is fast and helps you stay organised.*

Ordering Decimals to Find Median

QUESTION

At a school gardening club, Liam wrote these plant-growth measurements on cards:

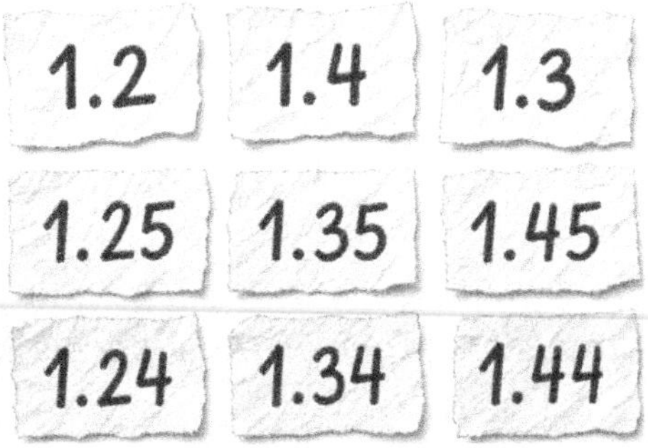

He places all the cards in order from **smallest to largest**.

Which number will be on the middle card?

LET'S UNDERSTAND THE QUESTION TOGETHER

We are trying to work out **which decimal will be in the centre** after all the cards are placed in order.

So, first, we need to arrange the numbers from **smallest to largest**.

Then we find the card that is **right in the middle**.

The middle card should have the same number of cards on each side.

WHAT WE KNOW

- ✓ The cards show these numbers:

  1.2, 1.4, 1.3, 1.25, 1.35, 1.45, 1.24, 1.34, 1.44

- ✓ The cards are placed in order from **smallest to largest.**
- ✓ There are **9 cards altogether.**

WHAT WE NEED TO FIND

We need to find **which number is on the middle card** after the decimals are arranged in order.

Our Plan

First, we will work out **which position is the middle**.

Because there are 9 numbers, the middle one is:

$$\frac{9+1}{2} = 5$$

So, we need to find the **5th number**.

Then we will put the decimals in order from **smallest to largest**.

To make that easier, we can write them with the same number of decimal places.

Step-by-Step Solution

Step 1: Find which position is the middle

There are 9 numbers.

So, the middle position is:

$$\frac{9+1}{2} = 5$$

This means we need to find the **5th number** in the ordered list.

Step 2: Write the decimals so they are easier to compare

Let's write each decimal to **2 decimal places**:

✓ $1.2 = 1.20$

✓ $1.4 = 1.40$

✓ $1.3 = 1.30$

✓ $1.25 = 1.25$

✓ $1.35 = 1.35$

✓ $1.45 = 1.45$

- ✓ 1.24 = 1.24
- ✓ 1.34 = 1.34
- ✓ 1.44 = 1.44

Step 3: Put the decimals in order from smallest to largest

Now we **order** them:

- ✓ 1.20
- ✓ 1.24
- ✓ 1.25
- ✓ 1.30
- ✓ 1.34
- ✓ 1.35
- ✓ 1.40
- ✓ 1.44
- ✓ 1.45

Step 4: Find the 5th number

Count along the ordered list:

1. 1.20
2. 1.24
3. 1.25
4. 1.30
5. **1.34**

So, the middle card is: **1.34**

Quick Check

There are **4 numbers smaller** than 1.34 and **4 numbers greater** than 1.34.

So, **1.34** is exactly in the **middle**.

Final Answer

The number on the middle card is **1.34**.

📌 Common Mistake

Do not choose the number that looks like it is in the middle of the picture. First, put all the numbers in order from smallest to largest. Then choose the middle number.

💡 Exam Tip

When ordering decimals, add extra zeros if needed. This helps you compare place value more easily.

Decimal Multiplication Place Value

QUESTION

Mika wanted to work out **0.4 × 23.5** using a digital cash register, but the decimal point key was not working.

So, she entered **4 × 235** instead, and the screen showed **940**.

Her mom challenged her to find the correct result **without multiplying directly**.

What is the correct value of **0.4 × 23.5**?

LET'S UNDERSTAND THE QUESTION TOGETHER

This question is about **multiplying decimals**.

Since the decimal point key was not working, Mika had to type the numbers without decimal points.
That made both numbers larger than they should be.

So, the answer on the screen is also too large.
We need to adjust it to find the real answer.

WHAT WE KNOW

- ✓ Mika wanted to find **0.4 × 23.5**
- ✓ She entered **4 × 235**
- ✓ The screen showed **940**
- ✓ **4** is **10 times** as large as **0.4**
- ✓ **235** is **10 times** as large as **23.5**

WHAT WE NEED TO FIND

We need to find the correct value of: **0.4 × 23.5**

Our Plan

We will compare the numbers Mika entered with the numbers she meant to enter.

- ✓ The first number became **10 times larger**
- ✓ The second number also became **10 times larger**

So, the answer became:

10 × 10 = 100 times larger

That means we must divide **940** by **100** to get the correct answer.

Step-by-Step Solution

Step 1: Work out how the numbers changed.

- ✓ **0.4** became **4**, so it became **10 times larger**
- ✓ **23.5** became **235**, so it also became **10 times larger**

Step 2: Work out how much larger the answer became.
If both numbers are multiplied by 10, the answer is multiplied by:

10 × 10 = 100

So, **940** is **100 times** too large.

Step 3: Divide to find the correct answer.

$940 \div 100 = 9.4$

So:

$0.4 \times 23.5 = 9.4$

Quick Check

We can check the answer using division.

If $0.4 \times 23.5 = 9.4$, then:

$9.4 \div 0.4 = 23.5$

Now work it out:

$9.4 \div 0.4 = 94 \div 4 = 23.5$

So, the multiplication is correct.

Final Answer

The correct value of 0.4×23.5 is **9.4**.

Common Mistake

*Do not divide 940 by 10. Both numbers were made 10 times larger, so the answer became **100 times larger**. Divide by **100**, not 10.*

Exam Tip

Use place value to keep track of the decimal point. Each time a number is made 10 times larger, its digits move one place to the left. That helps you see how the final answer must be adjusted.

Estimating Thousandths of a Unit

QUESTION

A kilogram is 1000 grams, so a **milligram** is one-thousandth of a gram.

In the same way, imagine a **millimonth**, which means **one-thousandth of a month**.

Which of the following is the closest estimate for a millimonth?

A. 3 seconds
B. 40 minutes
C. 45 minutes
D. 12 hours

LET'S UNDERSTAND THE QUESTION TOGETHER

This question is about breaking a big unit into **1000 equal parts**.

A month is a fairly long time, so one-thousandth of a month will be much smaller.
We need to work out about how long that smaller part would be.

The answer choices are given in different time units, so we need to estimate carefully.

WHAT WE KNOW

- ✓ A **millimonth** means **one-thousandth of a month.**
- ✓ 1 month is about **30 days.**
- ✓ We want to find **30 days ÷ 1000**
- ✓ Then we compare that result to the answer choices.

WHAT WE NEED TO FIND

We need to find which option is **closest to one-thousandth of a month**.

Our Plan

We will:

1. Estimate 1 month as **30 days.**
2. Change 30 days into a smaller unit we can divide more easily.
3. Divide by **1000**
4. Compare our result to the options.

This works because "milli" means **one-thousandth**.

Step-by-Step Solution

Step 1: Estimate the length of 1 month

A month is about **30 days**.

So:

$$\mathbf{1\ month \approx 30\ days}$$

Note: The symbol ≈ means **"is approximately equal to"** or **"is about"**. We use it here because not every month has exactly 30 days.

Step 2: Change days into hours

Each day has **24 hours**.

So:

$$\mathbf{30 \times 24 = 720\ hours}$$

That means:

$$\mathbf{30\ days = 720\ hours}$$

Step 3: Find one-thousandth of a month

A millimonth means **one-thousandth of a month.**

So, divide by **1000**:

$\mathbf{720 \div 1000 = 0.72\ hours}$

Step 4: Convert to minutes so we can compare with the options more easily

Our answer is **0.72 hours**, but some of the options are written in **minutes**.

Changing to minutes makes it easier to decide which choice is closest.

Since **1 hour = 60 minutes**:

0.72 × 60 = 43.2 minutes

Step 5: Compare with the answer choices

A millimonth is about **43.2 minutes**.

Now compare:

- **3 seconds** is far too short
- **40 minutes** is close
- **45 minutes** is even closer
- **12 hours** is far too long

So, the closest estimate is **45 minutes**.

Quick Check

We found that a millimonth is about **43.2 minutes**.
That is very close to **45 minutes**, so our answer is reasonable.

Final Answer

A millimonth is closest to **C. 45 minutes**.

Common Mistake

Do not divide the number of days by 1000 and stop there. After finding a tiny part of a month, convert it into hours or minutes so you can compare it with the answer choices.

Exam Tip

When working with time, convert to ***hours or minutes*** *before dividing. It often makes the comparison much easier.*

Fractions of a Circle (spinner)

QUESTION

A round game spinner is divided into 4 sectors.

- Two of the sectors have central angles labelled **a** and **b**, and **both are right angles**.
- One small sector is coloured **black** and takes up $\frac{1}{6}$ of the whole spinner.
- The remaining sector is coloured **grey**.

What fraction of the whole spinner is coloured grey?

LET'S UNDERSTAND THE QUESTION TOGETHER

This question is about parts of a whole circle.

A full circle is worth **1** or **360°**.

We already know:

- ✓ The black part is $\frac{1}{6}$ of the circle.
- ✓ Two other parts are right angles.

So, the grey part is the part that is **left over**.

WHAT WE KNOW

- ✓ A full circle = **360°**
- ✓ A right angle = **90°**
- ✓ Angle **a** is a right angle.
- ✓ Angle **b** is a right angle.
- ✓ The black sector is $\frac{1}{6}$ of the whole circle.

WHAT WE NEED TO FIND

We need to find the fraction of the circle that is coloured **grey**.

Our Plan

We will:

1. Change the black fraction into degrees
2. Add the known angles
3. Subtract from 360°
4. Convert the grey angle into a fraction of the whole circle

This works because all sectors in a circle must add up to one whole circle.

Step-by-Step Solution

Step 1: Find the angle of the black sector

The black sector is $\frac{1}{6}$ of the whole circle.

$$\frac{1}{6} \times 360^\circ = 60^\circ$$

So, the black sector is 60°.

Step 2: Add the two right angles

$$90^\circ + 90^\circ = 180^\circ$$

Now add the black sector:

$$180^\circ + 60^\circ = 240^\circ$$

So, the known parts add up to 240°.

Step 3: Find the grey sector

$$360^\circ - 240^\circ = 120^\circ$$

So, the grey sector is 120°.

Step 4: Write the grey sector as a fraction of the whole circle

$$\frac{120°}{360°} = \frac{1}{3}$$

Quick Check

Let's check the total:

$$90° + 90° + 60° + 120° = 360°$$

That makes a full circle, so our answer is correct.

Final Answer

The fraction of the spinner that is coloured **grey** is $\frac{1}{3}$.

Common Mistake

Mixing up which colour is the given fraction and which colour must be found.

Exam Tip

Two right angles make half a circle, so you can quickly think of them as $\frac{1}{2}$ *of the whole.*

Finding an Original Quantity (fractions)

QUESTION

At a school fair, Mia has a basket of raffle tickets.

She puts $\frac{1}{3}$ of the tickets into the first bag.

Then she puts $\frac{2}{5}$ of the remaining tickets into the second bag.

She has **16 tickets left** after that.

How many tickets did Mia have in the basket to begin with?

LET'S UNDERSTAND THE QUESTION TOGETHER

This question is about **fractions of a quantity**.

Mia starts with some tickets.
First, she puts part of them into one bag.
Then she puts part of the remaining tickets into another bag.
At the end, 16 tickets are left.

We need to find the number of tickets she had **at the beginning**.

WHAT WE KNOW

- ✓ Mia starts with some tickets.
- ✓ She puts $\frac{1}{3}$ of all the tickets into the first bag.
- ✓ That means $\frac{2}{3}$ of the tickets are left.
- ✓ She then puts $\frac{2}{5}$ of the remaining tickets into the second bag.
- ✓ After that, she has **16 tickets left**.

WHAT WE NEED TO FIND

We need to find **how many tickets Mia had in the basket to begin with.**

Our Plan

We will let the starting number of tickets be x.

Then we will:

- ✓ Find how many tickets are left after the first bag.
- ✓ Find how many go into the second bag.
- ✓ Work out how many are left after the second bag.
- ✓ Use the fact that this number is 16 to solve for x

This works because each fraction tells us what part of the tickets is being used or left over.

Step-by-Step Solution

Let the starting number of tickets be x.

Step 1: Find how many tickets go into the first bag

Mia puts $\frac{1}{3}$ of all the tickets into the first bag: $\frac{1}{3}x$

Step 2: Find how many tickets are left after the first bag

Subtract the tickets in the **first bag from the total**:

$$x - \frac{1}{3}x = \frac{2}{3}x$$

So, the number of tickets left after the first bag is: $\frac{2}{3}x$

Step 3: Find how many tickets go into the second bag

She puts $\frac{2}{5}$ of the remaining tickets into the second bag.

The remaining tickets are $\frac{2}{3}x$, so:

$$\frac{2}{5} \times \frac{2}{3}x = \frac{4}{15}x$$

So $\frac{4}{15}x$ tickets go into the second bag.

Step 4: Find how many tickets are left after the second bag

Subtract the tickets put into the second bag from the tickets remaining after the first bag:

$$\frac{2}{3}x - \frac{4}{15}x$$

The denominators are **3** and **15**. To subtract the fractions, they need the same denominator.

We change $\frac{2}{3}x$ into fifteenths because $3 \times 5 = 15$.
So we multiply both the numerator and denominator by 5:

$$\frac{2}{3}x = \frac{2 \times 5}{3 \times 5}x = \frac{10}{15}x$$

Now we can subtract:

$$\frac{10}{15}x - \frac{4}{15}x = \frac{6}{15}x = \frac{2}{5}x$$

So, the number of tickets left after the second bag is: $\frac{2}{5}x$

Step 5: Use the information given in the question

We are told that the number of tickets left is 16, so:

$$\frac{2}{5}x = 16$$

Step 6: Solve for x

Multiply both sides by $\frac{5}{2}$:

$x = 16 \times \frac{5}{2}$

$x = 40$

Quick Check

The three parts should add up to the starting number:

$$\frac{1}{3}x + \frac{4}{15}x + \frac{2}{5}x = x$$

Now replace $x$with **40**:

$\frac{1}{3}(40) + \frac{4}{15}(40) + \frac{2}{5}(40)$

$= \left(\frac{5}{15} + \frac{4}{15} + \frac{6}{15}\right) \times 40$

$= \frac{15}{15} \times 40$

$= 40$

So, all the parts add up to **40**, which means our answer makes sense.

Final Answer

Mia had **40** tickets in the basket to begin with.

Common Mistake

A common mistake is taking $\frac{2}{5}$ of the original number of tickets instead of taking $\frac{2}{5}$ of the

Exam Tip

*When fractions happen one after another, read carefully to see whether the next fraction is taken from the **original amount** or the **amount left over**.*

Fractions and Ratios in Groups

QUESTION

At a holiday reading program, there are **320 students**. **Five eighths** of the students are girls. Every student joins either the **Adventure Club** or the **Puzzle Club**.

Of the students in the **Adventure Club**, **3 out of every 10 are boys**. Also, **three quarters of all the boys** are in the **Puzzle Club**.

Answer these questions:

1. How many girls are in the Adventure Club?
2. What fraction of all students are in the Puzzle Club?

LET'S UNDERSTAND THE QUESTION TOGETHER

This question is about students being split into groups.

Some are **girls** and some are **boys**.
Some are in **Adventure Club** and some are in **Puzzle Club**.

We are given clues with fractions.
These clues help us work out the missing numbers.

The question wants us to find:

- ✓ How many **girls** are in Adventure Club.
- ✓ What **fraction of all students** are in Puzzle Club.

So, we need to read the clues carefully and understand how the groups fit together.

WHAT WE KNOW

- ✓ Total students = **320**
- ✓ $\frac{5}{8}$ of the students are girls.
- ✓ The rest are boys.
- ✓ Every student is in either **Adventure Club** or **Puzzle Club.**
- ✓ In the **Adventure Club,** $\frac{3}{10}$ are boys.
- ✓ $\frac{3}{4}$ of all the boys are in the **Puzzle Club.**

WHAT WE NEED TO FIND

We need to find:

- ✓ The number of **girls in the Adventure Club.**
- ✓ The **fraction of all students** who are in the Puzzle Club.

Our Plan

We will:

- ✓ Find the total number of **girls.**
- ✓ Find the total number of **boys.**
- ✓ Work out how many boys are in **Puzzle Club.**
- ✓ Then find how many boys are in **Adventure Club.**
- ✓ Use that clue to find the total number in **Adventure Club.**
- ✓ Find the number of girls in **Adventure Club.**
- ✓ Then find how many students are in **Puzzle Club.**
- ✓ Write that as a **fraction of all students.**

Step-by-Step Solution

Step 1: Find the number of girls

$$\frac{5}{8} \times 320 = 200$$

So, there are **200 girls.**

Step 2: Find the number of boys

$320 - 200 = 120$

So, there are **120 boys.**

Step 3: Find how many boys are in Puzzle Club

We know $\frac{3}{4}$ of all the boys are in Puzzle Club.

$$\frac{3}{4} \times 120 = 90$$

So, there are **90 boys in Puzzle Club.**

Step 4: Find how many boys are in Adventure Club

There are **120 boys** altogether.

90 boys are in Puzzle Club.

$120 - 90 = 30$

So, there are **30 boys in Adventure Club.**

Step 5: Find the total number in Adventure Club

If 30 boys are $\frac{3}{10}$ of Adventure Club, then:

$$\text{Adventure Club} = 30 \div \frac{3}{10} = 100$$

Step 6: Find the number of girls in Adventure Club

Adventure Club has **100 students.**

30 of them are boys.

$100 - 30 = 70$

So, there are **70 girls in Adventure Club.**

Step 7: Find the number of students in Puzzle Club

There are **320 students** altogether.
100 students are in Adventure Club.

$$320 - 100 = 220$$

So, there are **220 students in Puzzle Club**.

Step 8: Write this as a fraction of all students

$$\frac{220}{320}$$

Now simplify:

$$\frac{220}{320} = \frac{22}{32} = \frac{11}{16}$$

So, the fraction of all students in Puzzle Club is: $\frac{11}{16}$

Quick Check

Let's check:

- ✓ Girls = 200, Boys = 120
- ✓ Boys in Puzzle Club = **90**
- ✓ Boys in Adventure Club = **30**
- ✓ $\frac{3}{10}$ of 100 is 30, so that fits
- ✓ Girls in Adventure Club = **70**
- ✓ Puzzle Club = **320 - 100 = 220**

Everything matches the clues.

Final Answer

There are **70 girls in Adventure Club**, and $\frac{11}{16}$ of all students are in Puzzle Club.

Common Mistake

A common mistake is mixing up the ***whole group*** *and the* ***part of the group****. For example, 30 boys in Adventure Club is only one part of the club, not the whole club.*

Exam Tip

When a question has two kinds of groups, it helps to work with one clue at a time and fill in the numbers slowly.

Try to picture the groups clearly in your head. Ask, ***"Is this clue about boys and girls, or about the clubs?"*** *This can help you choose the right step.*

Algebra & Equations

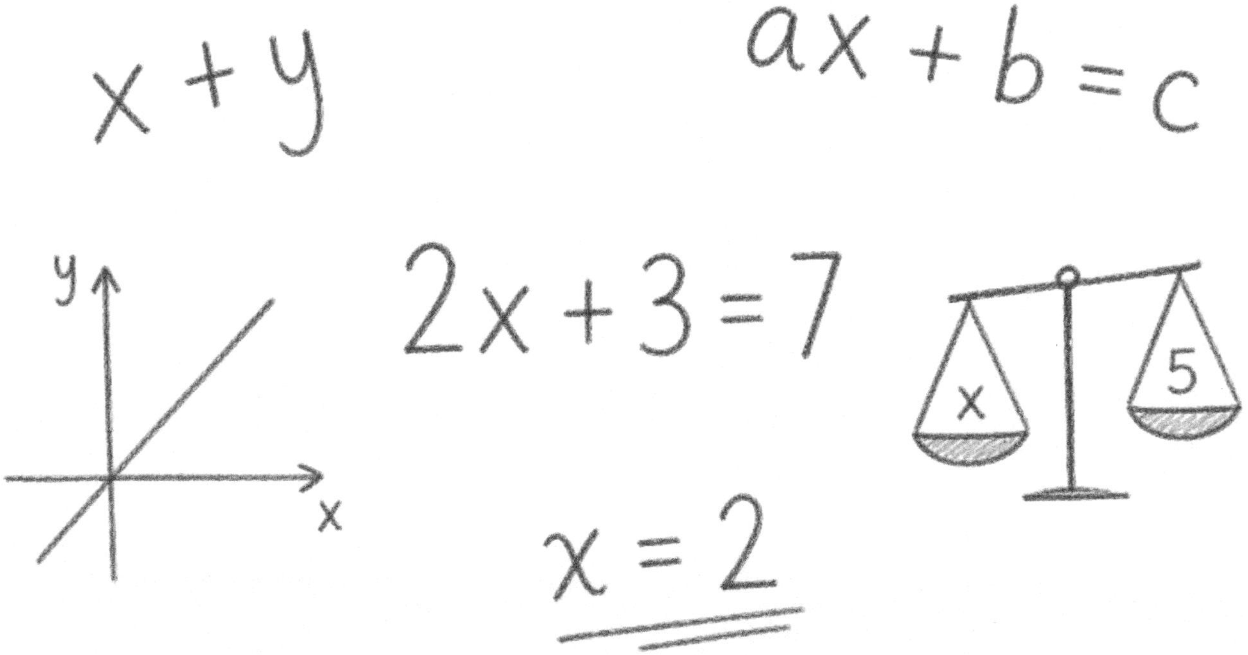

Finding a Missing Divisor

QUESTION

Mia says these two calculations are equal:

$$54 \div \square = 18 \div 3$$

What number should go in the box?

LET'S UNDERSTAND THE QUESTION TOGETHER

This question is about keeping **both sides** of an equation **equal**.

We will solve it by doing the same thing to both sides, one step at a time.

WHAT WE KNOW

- ✓ $54 \div \square = 18 \div 3$
- ✓ Both sides of the equation are equal.
- ✓ We can rewrite division as fractions to make the working easier to see.

WHAT WE NEED TO FIND

We need to find the number that should go in the box.

Our Plan

We will keep the equation balanced by doing the same thing to both sides.

1. First, we will rewrite the division equation in fraction form.
2. Then we will multiply both sides by 3.
3. Next, we will multiply both sides by the box to remove the denominator.
4. Finally, we will divide both sides by 18.

Step-by-Step Solution

We start with:

$$54 \div \square = 18 \div 3$$

Step 1: Write the division equation in fraction form

$$\frac{54}{\square} = \frac{18}{3}$$

Step 2: Multiply both sides by 3

$$\frac{54}{\square} \times 3 = \frac{18}{3} \times 3$$

$$\frac{162}{\square} = 18$$

Step 3: Multiply both sides by $\square$

$$\frac{162}{\square} \times \square = 18 \times \square$$

$$162 = 18\,\square$$

Step 4: Divide both sides by 18

$$\frac{162}{18} = \frac{18\square}{18}$$

$$9 = \square$$

So,

$$\square = 9$$

Quick Check

Let's check the answer:

$$54 \div 9 = 6$$

and

$$18 \div 3 = 6$$

Both sides are equal, so the answer is correct.

Final Answer

The number that should go in the box is **9**.

Common Mistake

A common mistake is changing only one side of the equation. Always do the same thing to both sides.

Exam Tip

When an equation has division, rewriting it as fractions can make the balancing steps much easier to see.

Finding a Missing Dividend

QUESTION

Liam says these two calculations are equal:

$$36 \div 9 = \triangle \div 5$$

What number should replace $\triangle$?

LET'S UNDERSTAND THE QUESTION TOGETHER

This question is about keeping **both sides of an equation equal.**

We need to find the value of the triangle so that both sides have the same answer.

WHAT WE KNOW

- ✓ $36 \div 9 = \triangle \div 5$
- ✓ Both sides of the equation are equal.
- ✓ We can rewrite division as fractions to make the working easier to see.

WHAT WE NEED TO FIND

We need to find the number that should replace $\triangle$.

Our Plan

We will keep the equation balanced by **doing the same thing to both sides**.

1. First, we will rewrite the division equation in fraction form.
2. Then we will multiply both sides by 9.
3. Next, we will multiply both sides by 5 to remove the denominator on the right side.

Step-by-Step Solution

We start with:

$$36 \div 9 = \triangle \div 5$$

Step 1: Write the division equation in fraction form

$$\frac{36}{9} = \frac{\triangle}{5}$$

Step 2: Multiply both sides by 9

$$\frac{36}{9} \times 9 = \frac{\triangle}{5} \times 9$$

$$36 = \frac{9\triangle}{5}$$

Step 3: Multiply both sides by 5

$$36 \times 5 = \frac{9\triangle}{5} \times 5$$

$$180 = 9\triangle$$

Step 4: Divide both sides by 9

$$\frac{180}{9} = \frac{9\triangle}{9}$$

$$20 = \triangle$$

So, $\triangle = 20$

Quick Check

Let's check the answer:

$$36 \div 9 = 4$$

And

$$20 \div 5 = 4$$

Both sides are equal, so the answer is correct.

Final Answer

The number that should replace △ is **20**.

Common Mistake

A common mistake is changing only one side of the equation. Always do the same thing to both sides.

Exam Tip

When an equation has division, rewriting it as fractions can make the balancing steps much easier to see.

Finding Missing Table Values (equations)

QUESTION

At a school canteen, the money collected from selling snacks was recorded in this table. Some amounts are missing.

	Drinks	Muffins	Row Total
Lunch	4.5	S	11.2
Recess	3.8	2.9	?
Column Total	8.3	?	T

1. What is the value of **S**?
2. What is the value of **T**?

LET'S UNDERSTAND THE QUESTION TOGETHER

This question is about adding numbers in a table.

- ✓ Each row total tells us how much the two numbers in that row add up to.
- ✓ Each column total tells us how much the two numbers in that column add up to.
- ✓ The overall total is the total of everything in the table.

So, we use addition and subtraction to fill in the missing parts.

WHAT WE KNOW

In the **Lunch** row:

- ✓ Drinks = **4.5**
- ✓ Muffins = **S**
- ✓ Row total = **11.2**

In the **Recess** row:

- ✓ Drinks = **3.8**
- ✓ Muffins = **2.9**

The **Drinks column total** is **8.3**

The overall total is **T**

WHAT WE NEED TO FIND

We need to find:

- ✓ **S**, the missing amount in the Lunch row.
- ✓ **T**, the total of all four numbers in the table.

Our Plan

We will use this idea:

- ✓ If we know the **row total** and one number in the row, we can find the missing number by **subtracting**
- ✓ Then we can add all the entries, or add the row totals, to find the **overall total**

This works because each total is just the sum of the numbers in that row or column.

Step-by-Step Solution

Step 1: Find the missing value **S**

Look at the **Lunch** row:

$$4.5 + S = 11.2$$

Subtract **4.5** from **11.2**:

$$S = 11.2 - 4.5$$
$$S = 6.7$$

So, the missing value is 6.7.

Step 2: Find the total for the Recess row

Add the two numbers in the Recess row:

$$3.8 + 2.9 = 6.7$$

So, the Recess row total is **6.7**.

Step 3: Find the overall total T

Add the two row totals:

$$T = 11.2 + 6.7$$
$$T = 17.9$$

So, the overall total is **17.9**.

Quick Check

Let's check our answers.

For the Lunch row:

$$4.5 + 6.7 = 11.2$$

Now check the overall total by adding all four inside numbers:

$$4.5 + 6.7 + 3.8 + 2.9 = 17.9$$

That matches our answer, so everything is correct.

Final Answer

The value of **S is 6.7**, and the value of **T is 17.9**.

Common Mistake

Some students forget to use subtraction when a part is missing from a total. Remember: if you know the total and one part, subtract to find the other part.

Exam Tip

Work row by row first. Once one missing number is found, the rest of the table usually becomes much easier.

Finding Missing Totals in a Table (equations)

QUESTION

A school library recorded the mass of books packed into two boxes on Friday and Saturday. Some numbers are missing.

	Story Books	Fact Books	Row Total
Friday	A	7.9	16.4
Saturday	6.8	B	C
Column Total	15.3	14.8	D

1. What is the value of **C**?
2. What is the value of **D**?

LET'S UNDERSTAND THE QUESTION TOGETHER

This one is a little more challenging because we cannot find both answers straight away.

We need to unlock the missing numbers in the correct order.

- ✓ First, we use a column total to find one missing value.
- ✓ Then we use that to find a row total.
- ✓ Then we use the row totals or column totals to find the grand total.

So, the question is really about careful step-by-step thinking.

WHAT WE KNOW

Friday row:

- ✓ Story Books = **A**
- ✓ Fact Books = **7.9**
- ✓ Row total = **16.4**

Saturday row:

- ✓ Story Books = **6.8**
- ✓ Fact Books = **B**
- ✓ Row total = **C**

Story Books column total = **15.3**

Fact Books column total = **14.8**

Overall total = **D**

WHAT WE NEED TO FIND

We need to find:

- ✓ **C**, the Saturday row total
- ✓ **D**, the overall total

Our Plan

We will start with the columns because both column totals are given.

That will help us find the missing inside numbers **A** and **B**.

Then we can add across to find **C**.

Finally, we can add the two row totals to find **D**.

This works because every row and column must balance correctly.

Step-by-Step Solution

Step 1: Find A using the Story Books column

The Story Books column total is 15.3.

$$A + 6.8 = 15.3$$

Subtract 6.8 from 15.3:

$$A = 15.3 - 6.8$$
$$A = 8.5$$

So, **A = 8.5**

Step 2: Find B using the Fact Books column

The Fact Books column total is 14.8.

$$7.9 + B = 14.8$$

Subtract 7.9 from 14.8:

$$B = 14.8 - 7.9$$
$$B = 6.9$$

So, **B = 6.9**

Step 3: Find C, the Saturday row total

Now add across the Saturday row:

$$6.8 + 6.9 = 13.7$$

So, the value of **C** is **13.7**.

Step 4: Find D, the overall total

Add the two row totals:

$16.4 + 13.7 = 30.1$

So, the value of **D** is **30.1**.

Quick Check

Check the Friday row:

$$8.5 + 7.9 = 16.4$$

Check the Saturday row:

$$6.8 + 6.9 = 13.7$$

Check the overall total using column totals:

$$15.3 + 14.8 = 30.1$$

That matches, so the answers make sense.

Final Answer

The value of **C is 13.7** and the value of **D is 30.1**.

📌 Common Mistake

Some students try to find the overall total too early. It is easier to find the missing inside numbers first.

💡 Exam Tip

In totals tables, begin with the row or column that has only one missing number. That usually unlocks the whole question.

Choosing Operations in an Equation

QUESTION

$16 - (8 \ ? \ 2) + 5 = 15$
What operation should replace the **?**

LET'S UNDERSTAND THE QUESTION TOGETHER

We can see a number sentence with one missing operation sign.

The sign goes between 8 and 2.

Our job is to work out **which operation makes the whole equation true**.

We are not finding a missing number. We are choosing the correct maths sign.

This could be plus, minus, multiply, or divide.

WHAT WE KNOW

The equation is:

$$16 - (8 \ ? \ 2) + 5 = 15$$

The missing symbol is an operation.

The bracket is worked out first.

WHAT WE NEED TO FIND

We need to find which operation between 8 and 2 makes the equation true.

Our Plan

We will simplify the left side a little first.

Then we will do the same operation to both sides to find what the bracket must equal.

After that, we will test which operation makes **8** ? **2** equal that number.

Step-by-Step Solution

Step 1: Write the full equation

$$16 - (8\ ?\ 2) + 5 = 15$$

Step 2: Simplify the left side a little

$$21 - (8\ ?\ 2) = 15$$

Step 3: Subtract 21 from both sides

$$21 - (8\ ?\ 2) - 21 = 15 - 21$$
$$-(8\ ?\ 2) = -6$$

Step 4: Multiply both sides by -1

$$8\ ?\ 2 = 6$$

Step 5: Test each operation

Try plus: $8 + 2 = 10$

Try minus: $8 - 2 = \mathbf{6}$

Try multiply: $8 \times 2 = 16$

Try divide: $\frac{8}{2} = 4$

Step 6: Choose the correct operation

The **operation** that makes the equation true is **minus.**

Quick Check

Put minus back into the equation:

$$16 - (8 - 2) + 5 = 15$$
$$16 - 6 + 5 = 15$$
$$10 + 5 = 15$$
$$15 = 15$$

Final Answer

The missing operation is **minus** (-).

📌 Common Mistake

Some students forget that the whole bracket is being subtracted.

💡 Exam Tip

Keep the equation balanced by doing the same thing to both sides.

Comparing Multiplicative Relationships

QUESTION

Liam, Zoe and Maya are collecting marbles.

Liam has **6 times** as many marbles as Zoe.

Maya has $\frac{1}{3}$ as many marbles as Zoe.

How many times as many marbles does Liam have as Maya?

LET'S UNDERSTAND THE QUESTION TOGETHER

We are trying to understand how Liam's number of marbles compares with Maya's number of marbles.

Both Liam's marbles and Maya's marbles are described using Zoe's marbles.

So, we need to use the information carefully to compare Liam and Maya.

WHAT WE KNOW

Liam has **6 times** as many marbles as Zoe.

Maya has $\frac{1}{3}$ as many marbles as Zoe.

We need to compare **Liam's marbles** with **Maya's marbles**.

WHAT WE NEED TO FIND

How many times as many marbles Liam has as Maya.

Our Plan

We will choose a number for Zoe's marbles that makes the maths easy.

Since Maya has $\frac{1}{3}$ as many marbles as Zoe, choosing **3** for Zoe is a smart idea because $\frac{1}{3}$ **of 3 is 1**.

Then we can work out:

- ✓ Maya's marbles.
- ✓ Liam's marbles.

After that, we will compare Liam's marbles with Maya's marbles by dividing.

Step-by-Step Solution

Since Maya has $\frac{1}{3}$ as many marbles as Zoe, let Zoe have **3 marbles**.

Step 1: Work out Maya's marbles

$$\frac{1}{3} \times 3 = 1$$

So, Maya has **1 marble**.

Step 2: Work out Liam's marbles

Liam has **6 times** as many marbles as Zoe.

$$6 \times 3 = 18$$

So, Liam has **18 marbles**.

Step 3: Compare Liam's marbles with Maya's marbles

Liam has **18 marbles** and Maya has **1 marble**.

$$18 \div 1 = 18$$

So, Liam has **18 times** as many marbles as Maya.

Quick Check

Let's check our answer.

- ✓ Zoe has **3 marbles.**
- ✓ Maya has **1 marble.**
- ✓ Liam has **18 marbles.**

Now check the comparison:

Maya's marbles $\times$ 18 = Liam's marbles

$1 \times 18 = 18$

Yes, that matches Liam's marbles, so the answer makes sense.

Final Answer

Liam has **18 times** as many marbles as Maya.

Common Mistake

*A common mistake is to compare the wrong two people. Make sure you compare **Liam with Maya**, not Liam with Zoe or Maya with Zoe.*

Exam Tip

*When a fraction is involved, choose a number that works nicely with the fraction. Since Maya has $\frac{1}{3}$ as many marbles as Zoe, choosing **3** makes the maths easier.*

Solving Cost Equations (unit prices)

QUESTION

At a school canteen, 3 sandwiches and 2 juices **cost the same as** 1 sandwich and 6 juices.

Each juice costs **$4**.

How much does **1 sandwich** cost?

LET'S UNDERSTAND THE QUESTION TOGETHER

This question is comparing **two canteen orders**.

Both orders cost exactly the same amount, even though they contain different numbers of sandwiches and juices.

We already know the cost of one item, and we need to use that information to find the cost of the other item.

WHAT WE KNOW

- ✓ **3 sandwiches and 2 juices** cost the same as **1 sandwich and 6 juices.**
- ✓ Each juice costs **$4**
- ✓ Both orders have the **same total cost.**

WHAT WE NEED TO FIND

We need to find the cost of **1 sandwich**.

Our Plan

Because the **total cost is equal**, we can use the **known price of the juice** to work out the price of the sandwich.

- ✓ First, we will write an equation to show that both orders cost the same.
- ✓ Then we will replace the juice cost with **$4**.
- ✓ After that, we will solve the equation to find the sandwich price.

Step-by-Step Solution

- ✓ Let the cost of **1 sandwich** be **S** dollars.
- ✓ Each juice costs **$4**.

Step 1: Write the equation

Since the two orders cost the same:

$$3 \text{ sandwiches} + 2 \text{ juices} = 1 \text{ sandwich} + 6 \text{ juices}$$

So:

$$3S + 2 \times 4 = 1S + 6 \times 4$$

Step 2: Work out the juice totals

$$3S + 8 = S + 24$$

Step 3: Subtract **S** from both sides:

$$2S + 8 = 24$$

Step 4: Remove the extra 8

Subtract **8** from both sides:

$$2S = 16$$

Step 5: Divide by 2

$S = 8$

So, one sandwich costs **$8**.

Quick Check

Let's test the answer.

Left side:

- ✓ 3 sandwiches = **3 × $8 = $24**
- ✓ 2 juices = **2 × $4 = $8**
- ✓ Total = **$32**

Right side:

- ✓ 1 sandwich = **$8**
- ✓ 6 juices = **6 × $4 = $24**
- ✓ Total = **$32**

Both sides cost the same, so the answer makes sense.

Final Answer

One sandwich costs **$8**.

Alternative Method

Since both orders cost the same, we can remove matching items from both sides.

3 sandwiches + 2 juices = 1 sandwich + 6 juices

Subtract 1 sandwich and 2 juices from both sides:

2 sandwiches = 4 juices

Each juice costs $4, so: 4 juices = $16

Therefore:

2 sandwiches = $16
1 sandwich = $8

Common Mistake

Some students forget that both sides must have the same ***total cost****, not the same number of items.*

Exam Tip

When two groups cost the same, look for items you can compare directly. Here, both sides include sandwiches and juices, so you can subtract the matching parts mentally to make the problem simpler before doing any equation.

Finding an Unknown in an Equation

QUESTION

$84 - 2(x + 5) = 47 + x$

What number does **x** represent?

LET'S UNDERSTAND THE QUESTION TOGETHER

The same missing number appears on both sides of the equation.

We need to find the one value that makes both sides equal.

To do that, we keep the equals sign the whole time and do the same thing to both sides.

WHAT WE KNOW

- ✓ The equation is:

  $$84 - 2(x + 5) = 47 + x$$

- ✓ The same number, **x**, appears on both sides.
- ✓ Both sides must be equal.

WHAT WE NEED TO FIND

We need to find the value of **x**.

Our Plan

We will simplify the left side first.

Then we will solve the equation by doing the same operation to both sides.

Step-by-Step Solution

Step 1: Expand the bracket.

$$84 - 2(x + 5) = 47 + x$$
$$84 - 2x - 10 = 47 + x$$

Step 2: Simplify the left side.

$$84 - 2x - 10 = 47 + x$$
$$74 - 2x = 47 + x$$

Step 3: Subtract 47 from both sides.

$$74 - 2x - 47 = 47 + x - 47$$
$$27 - 2x = x$$

Step 4: Add **2x** to both sides.

$$27 - 2x + 2x = x + 2x$$
$$27 = 3x$$

Step 5: Divide both sides by 3.

$$\frac{27}{3} = \frac{3x}{3}$$

So: $9 = x$

Then write: $x = 9$

Quick Check

Put **9** back into the equation:

Left side:

$$\mathbf{84 - 2(9 + 5) = 84 - 2(14) = 84 - 28 = 56}$$

Right side:

$$\mathbf{47 + 9 = 56}$$

Both sides are equal, so the answer is correct.

Final Answer

x = 9

Common Mistake

*Forgetting that the **2** multiplies both parts inside the bracket.*

Exam Tip

Keep the equation balanced by doing the same operation to both sides every time.

Brick Wall Number Patterns

QUESTION

In this **brick wall**, each brick is the sum of the two bricks directly below it.

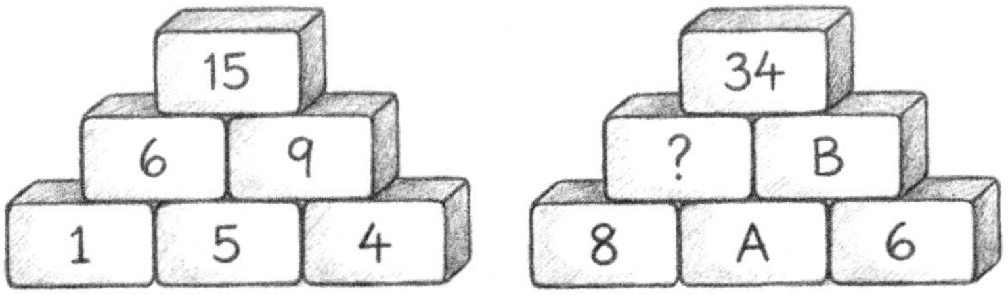

What is the value of **A + B**?

LET'S UNDERSTAND THE QUESTION TOGETHER

This question is about finding missing numbers in a brick wall.

Each brick above is made by adding the two bricks underneath it.

We need to use that pattern to work out **A** and **B**, then add them together.

WHAT WE KNOW

- ✓ Each brick is the sum of the two bricks directly below it.
- ✓ The top brick is **34**.
- ✓ The bottom row includes **8, A,** and **6**.
- ✓ One middle brick is **B**.

WHAT WE NEED TO FIND

We need to find: **A + B**

Our Plan

We will use the brick wall rule.

1. First, write the two middle bricks using the numbers below them.
2. Then use the top brick, which is **34**, to find **A**.
3. After that, find **B**.
4. Finally, add **A** and **B**.

Step-by-Step Solution

Step 1: Write the two middle bricks

The left middle brick is: $8 + A$

The right middle brick is: $A + 6$

Since the right middle brick is **B**, we know: $B = A + 6$

Step 2: Make an equation using the top brick

The top brick is the sum of the two middle bricks, so:

$(8 + A) + (A + 6) = 34$

Step 3: Simplify the left side

$8 + A + A + 6 = 34$
$14 + 2A = 34$

Step 4: Subtract 14 from both sides

$14 + 2A - 14 = 34 - 14$
$2A = 20$

Step 5: Divide both sides by 2

$$\frac{2A}{2} = \frac{20}{2}$$
$A = 10$

Step 6: Find B

$$B = A + 6$$
$$B = 10 + 6$$
$$B = 16$$

Step 7: Find A + B

$$A + B = 10 + 16$$
$$A + B = 26$$

Quick Check

Let's check the wall:

- ✓ Left middle brick: $8 + 10 = 18$
- ✓ Right middle brick: $10 + 6 = 16$
- ✓ Top brick: $18 + 16 = 34$

The top brick matches, so the answer is correct.

Final Answer

The value of **A + B is 26**.

Common Mistake

*Some students find **A** correctly but forget that the question asks for **A + B**, not just **A**. Always check exactly what the final question is asking.*

Exam Tip

Once you find one missing number, put it straight back into the wall. This makes the next step easier and helps you avoid mistakes.

Equivalent Fractions (fruit baskets)

QUESTION

At a fruit stall, the fraction of red apples in one basket is the same as the fraction of red apples in another basket.

$$\frac{3}{\square} = \frac{\square}{27}$$

What number should go in both boxes?

LET'S UNDERSTAND THE QUESTION TOGETHER

This question is about **equivalent fractions**.

Equivalent fractions are different-looking fractions that have the **same value**.

Here, the same missing number goes in **both boxes**. We need to find the number that makes the two fractions equal.

WHAT WE KNOW

The two fractions are equal.

The equation is:

$$\frac{3}{\square} = \frac{\square}{27}$$

The same number goes in both boxes.

WHAT WE NEED TO FIND

We need to find the number that makes both fractions equal.

Our Plan

We will let the missing number be $\boldsymbol{x}$.

Then the equation becomes:

$$\frac{3}{x} = \frac{x}{27}$$

To solve it, we can use **cross multiplication**.

This works because **when two fractions are equal, their cross products are equal too**.

Step-by-Step Solution

Let the missing number be x.

So:

$$\frac{3}{x} = \frac{x}{27}$$

Step 1: Cross multiply

$$3 \times 27 = x \times x$$
$$81 = x^2$$

Step 2: Find the value of x

We need the number that multiplies by itself to make 81.

$$9 \times 9 = 81$$

So: $x = 9$

Quick Check

Let's put 9 back into the fractions:

$$\frac{3}{9} = \frac{1}{3}$$

and

$$\frac{9}{27} = \frac{1}{3}$$

Both fractions are equal, so our answer is correct.

Final Answer

The number that goes in each box is **9**.

Common Mistake

Forgetting that the same box means the same number must be used both times.

Exam Tip

When the missing number appears in both fractions, cross multiplication is usually the fastest way to solve it.

One-Step Linear Equation (puzzle)

QUESTION

A number puzzle uses a **basket**, a **bag**, and a **tag**.

- ✓ The number in the **basket** is multiplied by 4.
- ✓ The number in the **bag** is multiplied by 3.
- ✓ The difference between these two answers is written on the **tag**.

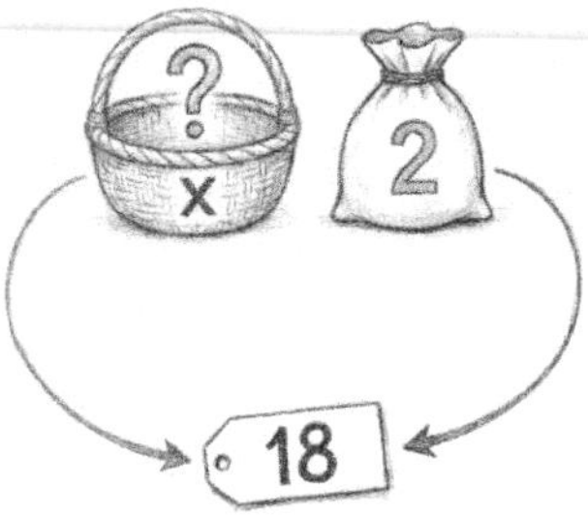

The number in the basket is greater than 2.

What number should go in the basket?

LET'S UNDERSTAND THE QUESTION TOGETHER

This question gives us a picture with three objects: a basket, a bag, and a tag.

The basket has a missing number, the bag has the number **2**, and the tag has the number **18**.

We are told that the numbers in the basket and bag are used in different calculations, and the tag shows the difference between the two results.

So, this question is really about understanding how the three objects are connected and what the missing number in the basket must be.

WHAT WE KNOW

- ✓ The **basket** shows **?**
- ✓ The **bag** shows **2**
- ✓ The **tag** shows **18**
- ✓ The number in the **basket** is multiplied by **4**
- ✓ The number in the **bag** is multiplied by **3**
- ✓ The **tag** shows the **difference** between the two answers
- ✓ The number in the basket is greater than **2**

WHAT WE NEED TO FIND

We need to find the missing number in the **basket**.

Our Plan

We will:

1. Find the result for the **bag**.
2. Use the number on the **tag** to work out the basket's result.
3. Divide by 4 to find the starting number in the **basket**.

This works because the tag shows the difference between the two multiplied results.

Step-by-Step Solution

Step 1: Find the bag's result

The bag has **2**.

It is multiplied by **3**:

$2 \times 3 = 6$

So, the bag's result is **6**.

Step 2: Use the tag

The tag shows the difference, which is **18**.

Since the number in the basket is greater than 2, the basket's result must be greater than the bag's result.

So:

$$\textbf{basket result} - 6 = 18$$

Add 6 to both sides:

$$\textbf{basket result} = 24$$

So, the basket's multiplied result is **24**.

Step 3: Find the basket number

The basket number was multiplied by **4**.

So:

$$\textbf{basket number} \times 4 = 24$$

Divide both sides by 4:

$$\textbf{basket number} = 24 \div 4$$

$$\textbf{basket number} = 6$$

Quick Check

Let's check:

- ✓ Basket: $6 \times 4 = 24$
- ✓ Bag: $2 \times 3 = 6$
- ✓ Difference: $24 - 6 = 18$

The tag shows **18**, so our answer is correct.

Final Answer

The missing number in the basket is **6**.

Alternative Method (Using an Equation)

Let the basket number be x.

$$4x - 3(2) = 18$$
$$4x - 6 = 18$$
$$4x = 24$$
$$x = 6$$

Common Mistake

A common mistake is forgetting to work backwards. The tag does not tell us the basket's answer directly, so we must first use the bag's result and the difference.

Exam Tip

*Underline the clue that tells you what the final number means. Words like **difference, total,** or **product** tell you which maths operation to use.*

Working Out Prices from Clues

QUESTION

At the school canteen, the canteen lady accidentally spilled water all over the price board, so some of the prices washed away!

The board showed these items:

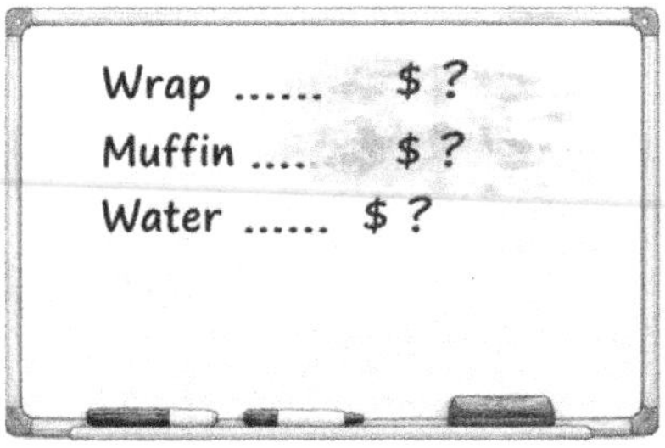

Luckily, some students still remembered these totals:

- ✓ A **wrap and a water** cost **$6.40** altogether.
- ✓ A **muffin and a water** cost **$4.10** altogether.
- ✓ A **wrap, a muffin and a water** cost **$7.85** altogether.

How much do a **wrap and a muffin** cost altogether?

LET'S UNDERSTAND THE QUESTION TOGETHER

This is a price mystery.

The water made the price board messy, but we still have enough clues to solve it.
We are not trying to find each item one by one straight away.
We only need the total cost of the **wrap and the muffin**.

WHAT WE KNOW

- ✓ Wrap + Water = **$6.40**
- ✓ Muffin + Water = **$4.10**
- ✓ Wrap + Muffin + Water = **$7.85**

WHAT WE NEED TO FIND

We need to find:

- ✓ **Wrap + Muffin**

Our Plan

We will first find the cost of the **water**.

Why? Because the total for **wrap + muffin + water** includes the water.
Once we know the water price, we can subtract it from the 3-item total.

That will leave us with just:

- ✓ **wrap + muffin**

Step-by-Step Solution

We use letters to represent the unknown amounts:

- ✓ Wrap = W
- ✓ Muffin = M
- ✓ Water = T

We know:

$$W + T = 6.40$$
$$M + T = 4.10$$
$$W + M + T = 7.85$$

Step 1: Add the first two totals

We add these two because they both include **water**.

$$(W + T) + (M + T) = 6.40 + 4.10$$

$$W + M + 2T = 10.50$$

Step 2: Use the total of all 3 items

We also know:

$$W + M + T = 7.85$$

Now subtract:

$$(W + M + 2T) - (W + M + T) = 10.50 - 7.85$$

$$T = 2.65$$

So, the **water costs $2.65**.

Step 3: Find wrap + muffin

Now use:

$$W + M + T = 7.85$$

Substitute $T = 2.65$:

$$W + M + 2.65 = 7.85$$

Subtract 2.65 from both sides:

$$W + M = 7.85 - 2.65$$
$$W + M = 5.20$$

Quick Check

Let's check:

- ✓ Wrap + Muffin = **$5.20**
- ✓ Water = **$2.65**

So:

$$5.20 + 2.65 = 7.85$$

That matches the total, so our answer is correct.

Final Answer

A **wrap and a muffin cost $5.20 altogether**.

Common Mistake

Forgetting that when we add the first two totals, the water is counted twice.

Exam Tip

When one item appears in lots of totals, try finding that item first. It often makes the rest of the question much easier.

Finding Full Weight from Fractions

QUESTION

The diagram shows **two identical** water pitchers on digital scales, with their masses shown.

1/3 Full　　2/3 Full

What is the **total mass** of the pitcher and water when the pitcher is **completely full**?

LET'S UNDERSTAND THE QUESTION TOGETHER

This question is about separating two parts:

- ✓ The mass of the **empty pitcher.**
- ✓ The mass of the **water.**

The **pitcher** always **weighs the same.**

Only the amount of water changes.

So, we can use algebra to represent each situation clearly and solve it step by step.

WHAT WE KNOW

- ✓ One-third full pitcher + water = **90 g**
- ✓ Two-thirds full pitcher + water = **150 g**
- ✓ The empty pitcher has a fixed mass.
- ✓ A full amount of water can be split into equal thirds.

WHAT WE NEED TO FIND

We need to find the **total mass of the pitcher and water when the pitcher is completely full**.

Our Plan

We will use algebra.

Let:

- ✓ $P =$ mass of the empty pitcher
- ✓ $W =$ mass of the water when the pitcher is completely full

Then we can write two equations:

$$P + \frac{1}{3}W = 90$$
$$P + \frac{2}{3}W = 150$$

After that, we subtract the equations to find W, then work out P, and finally add them together.

Step-by-Step Solution

Step 1: We use letters to represent the unknown amounts.

- ✓ $P =$ mass of the empty pitcher
- ✓ $W =$ mass of the water when the pitcher is completely full

Now we can write the two situations as equations.

From the question:

$$P + \frac{1}{3}W = 90$$
$$P + \frac{2}{3}W = 150$$

Step 2: Subtract the first equation from the second

$$\left(P+\frac{2}{3}W\right)-\left(P+\frac{1}{3}W\right)=150-90$$

Now simplify:

$$P-P+\frac{2}{3}W-\frac{1}{3}W=60$$
$$\frac{1}{3}W=60$$

Step 3: Find W

If one-third of W is 60, then the full amount is:

$$W=60\times 3$$
$$W=180$$

So, the water in a completely full pitcher weighs **180 g**.

Step 4: Find P

Substitute $W=180$into the first equation:

$$P+\frac{1}{3}(180)=90$$
$$P+60=90$$
$$P=30$$

So, the empty pitcher weighs **30 g**.

Step 5: Find the total mass when the pitcher is full

$$P+W=30+180$$
$$P+W=210$$

Quick Check

Let's test both clues:

- ✓ One-third full:

  $$30 + \frac{1}{3}(180) = 30 + 60 = 90$$

- ✓ Two-thirds full:

  $$30 + \frac{2}{3}(180) = 30 + 120 = 150$$

So, our answer makes sense.

Final Answer

When the pitcher is completely full, the pitcher and water together weigh **210 g**.

Common Mistake

Stopping at $W = 180$. That is only the mass of the water. The question asks for the mass of the ***full pitcher and water together****.*

Exam Tip

When two algebra equations both have the same variable, subtracting them can be a quick way to eliminate one variable and make the problem much easier.

Working Out Meal Deal Prices

QUESTION

At a school canteen, three different meal deals have different prices:

What is the price of the **salad**?

LET'S UNDERSTAND THE QUESTION TOGETHER

This question is about finding unknown numbers using clues.

We have three food items, but we do not know their prices yet.
The clues tell us how the sandwich price and juice price are connected.

Once we figure out those two prices, we can add them together to find the price of the salad.

So, this is really a puzzle about using number sentences carefully.

WHAT WE KNOW

- ✓ The **sandwich** costs **3 times** the price of the **juice**.
- ✓ The **sandwich** costs **$18 more** than the **juice**.
- ✓ **Sandwich + juice = salad**

WHAT WE NEED TO FIND

We need to find the price of the **salad**.

Our Plan

We will **first** find the prices of the **sandwich** and the juice.

Why does this work?
Because the first two clues both describe the same sandwich price in different ways.

So, we can compare those clues to **work out the value of the juice**.
Then we can use the last clue to find the **salad price**.

Step-by-Step Solution

Step 1: Let the price of the juice be $j$dollars.

Then the sandwich costs: $\mathbf{3j}$

Step 2: Write an equation using the second clue.

We also know the sandwich costs 18dollars more than the juice, so:

$\mathbf{3j = j + 18}$

Step 3: Solve the equation.

Subtract $\mathbf{j}$ from both sides:

$\mathbf{2j = 18}$

Divide both sides by **2**:

$\mathbf{j = 9}$

So, the **juice costs $9**.

Step 4: Find the price of the sandwich.

$\mathbf{3 \times 9 = 27}$

So, the **sandwich costs $27**.

Step 5: Find the price of the salad.

Salad = sandwich + juice

$\mathbf{27 + 9 = 36}$

So, the **salad costs $36**.

Quick Check

Let's test the clues:

- ✓ Is the sandwich 3 times the juice? $3 \times 9 = 27$
- ✓ Is the sandwich $18 more than the juice? $9 + 18 = 27$
- ✓ Is the salad the total of sandwich and juice? $27 + 9 = 36$

Everything matches.

Final Answer

The **salad costs $36**

📌 **Common Mistake**

*The clues compare the **item prices**, not the numbers by themselves.*

💡 **Exam Tip**

If two clues describe the same value, set them equal to solve quickly.

Finding the Starting Amount

QUESTION

Mika packed some fruit bars in a box for a picnic.

First, she gave **one-third** of the fruit bars to her brother.

Then she gave **18** of the fruit bars that were left to her cousin.

After that, there were **30** fruit bars still left in the box, and Mika kept these for herself.

How many fruit bars were in the box at the start?

LET'S UNDERSTAND THE QUESTION TOGETHER

This question is about sharing a starting amount in two steps.

- ✓ First, Mika gives away **one-third** of the fruit bars.
- ✓ Then she gives away **18 more**.
- ✓ After all that, **30** are left.

We need to find the number she had **at the beginning**.

WHAT WE KNOW

- ✓ Mika started with some fruit bars.
- ✓ She gave **one-third** of them to her brother.
- ✓ Then she gave **18** fruit bars to her cousin.
- ✓ She had **30** fruit bars left at the end.

WHAT WE NEED TO FIND

We need to find **how many fruit bars Mika had at the start.**

Our Plan

We will use **algebra**.

We can let the starting number be $\boldsymbol{x}$, then carefully follow each step in the question.

This helps us turn the word problem into an equation we can solve.

Step-by-Step Solution

Step 1: Let the starting number be x.

Step 2: After giving **one-third** to her brother, Mika had:

$$x - \frac{x}{3} = \frac{2x}{3}$$

Step 3: Then she gave **18** to her cousin, leaving **30**:

$$\frac{2x}{3} - 18 = 30$$

Step 4: Add **18** to both sides:

$$\frac{2x}{3} = 48$$

Step 5: Multiply both sides by $\frac{3}{2}$:

$$x = 48 \times \frac{3}{2}$$
$$x = 72$$

So, Mika started with **72 fruit bars.**

Quick Check

Let's check:

- ✓ Start with **72**
- ✓ One-third of 72 is **24**
- ✓ $72 - 24 = 48$
- ✓ $48 - 18 = 30$

That matches the question perfectly.

Final Answer

Mika had **72 fruit bars in the box at the start**.

Common Mistake

Making errors when multiplying by a fraction, especially forgetting to flip $\frac{2}{3}$ *to* $\frac{3}{2}$.

Exam Tip

When a problem starts with "some number," using x can make the steps much easier to follow.

System of Linear Equations (fundraising context)

QUESTION

Four classes raised **$800** altogether for charity.

- ✓ Class B raised **twice as much** as Class C.
- ✓ Class A raised **3 times as much** as Class C.
- ✓ Class D raised **$140**.

How much money did **Class C** raise?

LET'S UNDERSTAND THE QUESTION TOGETHER

This question is about splitting a total amount of money between four classes.

Some classes raised amounts that are connected to **Class C**. That means Class C is a good place to start.

Once we write the other class amounts using Class C, we can add everything together and solve it.

WHAT WE KNOW

- ✓ Total money raised = **$800**
- ✓ Class B raised **2 times** Class C
- ✓ Class A raised **3 times** Class C
- ✓ Class D raised **$140**

WHAT WE NEED TO FIND

We need to find **how much Class C raised**.

Our Plan

We will let Class C's amount be one unknown value.

Then we will:

- ✓ Write Class A and Class B in terms of Class C.
- ✓ Add all four class amounts.
- ✓ Make an equation.
- ✓ Solve it.

This works because all the money together must equal **$800**.

Step-by-Step Solution

Step 1: Let the amount raised by **Class C** be **$x**.

Then:

- ✓ Class A raised **$3x**
- ✓ Class B raised **$2x**
- ✓ Class C raised **$x**
- ✓ Class D raised **$140**

So:

$$3x + 2x + x + 140 = 800$$

Step 2: Solve the equation.

Combine like terms:

$$6x + 140 = 800$$

Subtract 140 from both sides:

$$6x = 660$$

Divide by 6:

$$x = 110$$

So, **Class C raised $110**.

Quick Check

Let's check:

- ✓ Class C = **$110**
- ✓ Class B = **$220**
- ✓ Class A = **$330**
- ✓ Class D = **$140**

Now add them:

$$\mathbf{330 + 220 + 110 + 140 = 800}$$

That matches the total, so the answer is correct.

Final Answer

Class C raised **$110.**

Common Mistake

A common mistake is forgetting to include Class C itself when adding the amounts. Remember, the total is made from $\boldsymbol{A + B + C + D}$.

Exam Tip

When several amounts are compared to one class, make that class the unknown first. It usually makes the equation much simpler.

Linear Equations (charity totals)

QUESTION

Four classes raised **$700** altogether for charity.

- ✓ Class **A** raised **twice** as much as Class C
- ✓ Class **B** raised **half** as much as Class C
- ✓ Class **D** raised **$140**

How much did **Class C** raise?

LET'S UNDERSTAND THE QUESTION TOGETHER

This question is about sharing a total amount of money between four classes.

Some of the class amounts are linked to **Class C**. One class raised **twice as much**, and another raised **half as much**.

That makes the question a little trickier, we need to work carefully because one amount is written as a fraction of Class C.

WHAT WE KNOW

- ✓ Total money raised = **$700**
- ✓ Class A raised **twice as much as Class C**
- ✓ Class B raised **half as much as Class C**
- ✓ Class D raised **$140**

WHAT WE NEED TO FIND

We need to find **how much Class C raised**.

Our Plan

We will let the amount raised by **Class C** be $\boldsymbol{x}$.

Then we can write the amounts for the other classes using $\boldsymbol{x}$.

After that, we add all four class amounts and make an equation equal to **$700**.

Then we solve the equation step by step.

Step-by-Step Solution

Step 1: Let the amount raised by **Class C** be x.

Then:

- ✓ Class A $= 2x$
- ✓ Class B $= \frac{x}{2}$
- ✓ Class C $= x$
- ✓ Class D $= \mathbf{140}$

So, the total is:

$$2x + \frac{x}{2} + x + 140 = 700$$

Step 2: Rewrite the $\boldsymbol{x}$ terms so they all have the same denominator.

We do this because $\frac{x}{2}$ is in **halves**, so it is easier to add all the $\boldsymbol{x}$ terms if they are all written in halves.

$$\frac{4x}{2} + \frac{x}{2} + \frac{2x}{2} + 140 = 700$$

Now we can combine the $x$terms:

$$\frac{7x}{2} + 140 = 700$$

Step 3: Solve the equation.

Subtract 140 from both sides:

$$\frac{7x}{2} = 560$$

Multiply both sides by 2:

$$7x = 1120$$

Divide by 7:

$$x = 160$$

So, **Class C raised $160**.

Quick Check

Let's check:

- ✓ Class C = **$160**
- ✓ Class A = **$320**
- ✓ Class B = **$80**
- ✓ Class D = **$140**

Now add them:

$$320 + 80 + 160 + 140 = 700$$

The total matches, so the answer is correct.

Final Answer

Class C raised $160.

Common Mistake

A common mistake is forgetting to rewrite all the x terms with the same denominator before combining them.

Exam Tip

When one term is a fraction, rewrite the other like terms to match it. This makes the algebra much easier to handle.

Compound Inequalities (books per box)

QUESTION

A school library is putting books into boxes.
Each box has the same number of books.

If the number of books in one box is multiplied by **5**, the answer is **less than 65**.

If the number of books in one box is multiplied by **8**, the answer is **more than 80**.

There is more than one possible number of books in each box.

What is the **sum of all the possible** numbers of books in one box?

LET'S UNDERSTAND THE QUESTION TOGETHER

This question is giving us two clues about how many books could be in each box.

1. One clue tells us the number is not too big.
2. The other clue tells us the number is not too small.

We need to find all the whole numbers that fit both clues.

Then we add them together.

WHAT WE KNOW

- ✓ The books are put into boxes.
- ✓ Each box has the **same number of books.**
- ✓ If the number is multiplied by 5, the answer is **less than 65.**
- ✓ If the number is multiplied by 8, the answer is **more than 80.**
- ✓ There is **more than one** possible number.
- ✓ The number of books is a **whole number.**

WHAT WE NEED TO FIND

We need to find the **sum of all possible numbers of books** in one box.

Our Plan

We will use each clue to make a number rule.

Then we will find the whole numbers that fit both rules.

After that, we will add those numbers.

This works because the correct answers must match both clues.

Step-by-Step Solution

Let the number of books in one box be n.

Step 1: Use the first clue

The question says:

$$5n < 65$$

Divide both sides by 5:

$$n < 13$$

So, the number of books must be less than 13.

Step 2: Use the second clue

The question also says:

$$8n > 80$$

Divide both sides by 8:

$$n > 10$$

So, the number of books must be more than 10.

Step 3: Find the whole numbers that fit both clues

The number must be:

- ✓ More than 10
- ✓ Less than 13

The whole numbers are: 11, 12

Step 4: Add the possible numbers

$$11 + 12 = 23$$

Quick Check

Check 11:

- ✓ $5 \times 11 = 55$, which is less than 65
- ✓ $8 \times 11 = 88$, which is more than 80

Check 12:

- ✓ $5 \times 12 = 60$, which is less than 65
- ✓ $8 \times 12 = 96$, which is more than 80

Both numbers work.

Final Answer

The sum of all the possible numbers of books in one box is **23**.

Common Mistake

Including 10 or 13. They do not work because the question says more than 10 and less than 13.

Exam Tip

Find the lower limit and the upper limit first. Then list the whole numbers in between.

Digit Reversal (two-digit numbers)

QUESTION

A two-digit locker code is written as a whole number between 10 and 99.
When the digits of the code are swapped, the new number is 54 greater than the original number.
What are the possible ones digits of the original locker code?

LET'S UNDERSTAND THE QUESTION TOGETHER

This question is about a **two-digit number** and what happens when we **swap its digits**.

For example, if a number was 23, swapping the digits would make 32.

Sometimes the new number is bigger. We need to work out which original number could do that and then find its **ones digit**.

WHAT WE KNOW

- ✓ The number has **2 digits**.
- ✓ It is between **10 and 99**.
- ✓ When the digits are reversed, the new number is **54 more** than the original number.
- ✓ We need the **ones digit** of the original number.

WHAT WE NEED TO FIND

We need to find the **last digit** (ones digit) of the original two-digit number.

Our Plan

We will represent the two-digit number using:

- ✓ A **tens digit.**
- ✓ A **ones digit.**

Then we will write both numbers algebraically and compare them.

This works well because place value is very important here.

Step-by-Step Solution

Step 1: Represent the original number

The number has two digits:

- ✓ The **tens digit** is **a**
- ✓ The **ones digit** is **b**

So, the original number is: $\mathbf{10a + b}$

Step 2: Write the reversed number

When the digits are swapped, the new number becomes: $\mathbf{10b + a}$

Step 3: Use the information from the question

We are told that the reversed number is **54 more** than the original number.

So, we can write:

$$\textbf{Reversed number} = \textbf{Original number} + 54$$

$$10b + a = 10a + b + 54$$

Step 4: Simplify the equation

Take like terms together:

$$10b - b + a - 10a = 54$$

$$9b - 9a = 54$$

Step 5: Factor and solve

Factor out **9**:

$$9(b - a) = 54$$

Now divide both sides by **9**:

$$b - a = 6$$

This means the **ones digit is 6 more than the tens digit**.

Step 6: Find possible digit pairs

Since digits must be between **0 and 9**, we test possible values:

- ✓ if **a = 1**, then **b = 7**
- ✓ if **a = 2**, then **b = 8**
- ✓ if **a = 3**, then **b = 9**

So, the possible original numbers are:

- ✓ **17**
- ✓ **28**
- ✓ **39**

Step 7: Find the possible ones digits

The ones digits of these numbers are:

- ✓ **7**
- ✓ **8**
- ✓ **9**

Step 8: Check the possibilities

Let's test them:

- **17 → 71**, and **71 - 17 = 54**
- **28 → 82**, and **82 - 28 = 54**
- **39 → 93**, and **93 - 39 = 54**

Step 9: State the result

All of them work. So, there is **more than one possible original number**.

There are multiple possible values for the ones digit.

The possible ones digits are:

7, 8, or 9

Quick Check

Check one example:

- ✓ Original number: **28**
- ✓ Reversed number: **82**
- ✓ Difference: **82 - 28 = 54**

It works.

Final Answer

The original code could have a ones digit of **7, 8, or 9**.

Common Mistake

Forgetting that the tens digit and ones digit have different values. For example, in 28, the 2 means 20, not just 2.

Exam Tip

*When digits are reversed in a two-digit number, the difference is always a multiple of **9**. That can help you solve these questions faster.*

Division with Remainder (word problem)

QUESTION

A toy shop packed **marbles** into two jars.

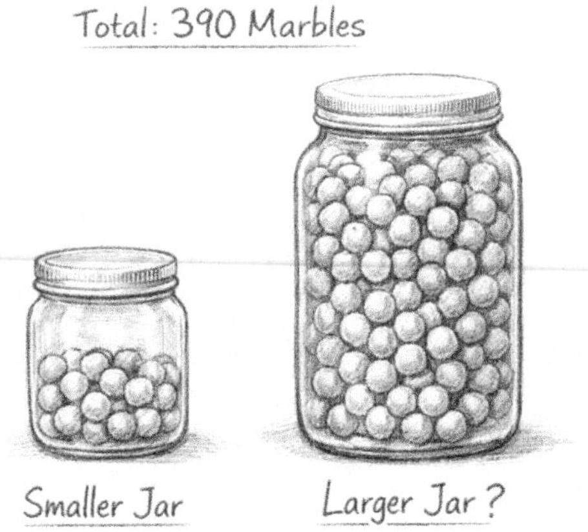

If the number of marbles in the **larger jar** is **divided** by the number of marbles in the **smaller jar**, the quotient is **2** and the remainder is **30**.

How many marbles are in the larger jar?

LET'S UNDERSTAND THE QUESTION TOGETHER

This question is about **sharing marbles between two jars.**

We know the total number of marbles in both jars is **390**.
We also know the larger jar has a number of marbles that is **2 times** the smaller jar, with **30 extra marbles left over**.

So, we need to use both clues to work out how many marbles are in the larger jar.

WHAT WE KNOW

- ✓ There are **390 marbles** altogether
- ✓ The marbles are placed in **two jars**
- ✓ Larger jar ÷ smaller jar gives:
    - Quotient = **2**
    - Remainder = **30**

WHAT WE NEED TO FIND

We need to find **how many marbles are in the larger jar**.

Our Plan

We will use the clue about the quotient and remainder to write the larger jar amount.

Since the quotient is **2** and the remainder is **30**:

larger jar = 2 × smaller jar + 30

Then we will add the smaller jar and larger jar together, because the total is **390 marbles**.

Step-by-Step Solution

Step 1: Let the smaller jar have some number of marbles

We will call the number of marbles in the smaller jar **s**.

So: **smaller jar** = s

Step 2: Write the larger jar in terms of the smaller jar

We are told that:

- ✓ The quotient is **2**
- ✓ The remainder is **30**

This means the larger jar has:

- ✓ **2 groups of the smaller jar**
- ✓ and **30 extra marbles**

So:

larger jar $= \mathbf{2 \times s + 30}$

Step 3: Use the total number of marbles

The two jars together have **390 marbles.**

So:

Small jar + Large jar $= 390$

$s + (2 \times s + 30) = 390$

Step 4: Find the value of $\boldsymbol{s}$

$$3s = 360$$

Divide both sides by 3:

$$s = 120$$

So, the smaller jar has **120 marbles.**

Step 5: Find the larger jar

Start with the formula:

$$\text{larger jar} = 2 \times s + 30$$

Substitute $s = 120$:

$$\text{larger jar} = 2 \times 120 + 30 = 240 + 30 = 270$$

So, the larger jar has **270 marbles.**

Quick Check

Check the total:

$$120 + 270 = 390$$ ✓

Check the division:

$$270 \div 120 = 2 \text{ remainder } 30$$ ✓

Final Answer

The larger jar has **270 marbles**.

Common Mistake

Forgetting to add the remainder. The larger jar is not just $2\times$ *the smaller jar. It is* $2\times$ ***the smaller jar + 30****.*

Exam Tip

In quotient and remainder questions, remember:

Here's what each word means:

- ✓ ***Dividend*** *→ the number you are dividing*
  In this question, it is the ***larger number***
- ✓ ***Divisor*** *→ the number you are dividing by*
  This is the ***smaller number***
- ✓ ***Quotient*** *→ the answer you get from dividing*
  Here, it is ***2***
- ✓ ***Remainder*** *→ what is left over after dividing*
  Here, it is ***30***

Geometry & Measurement

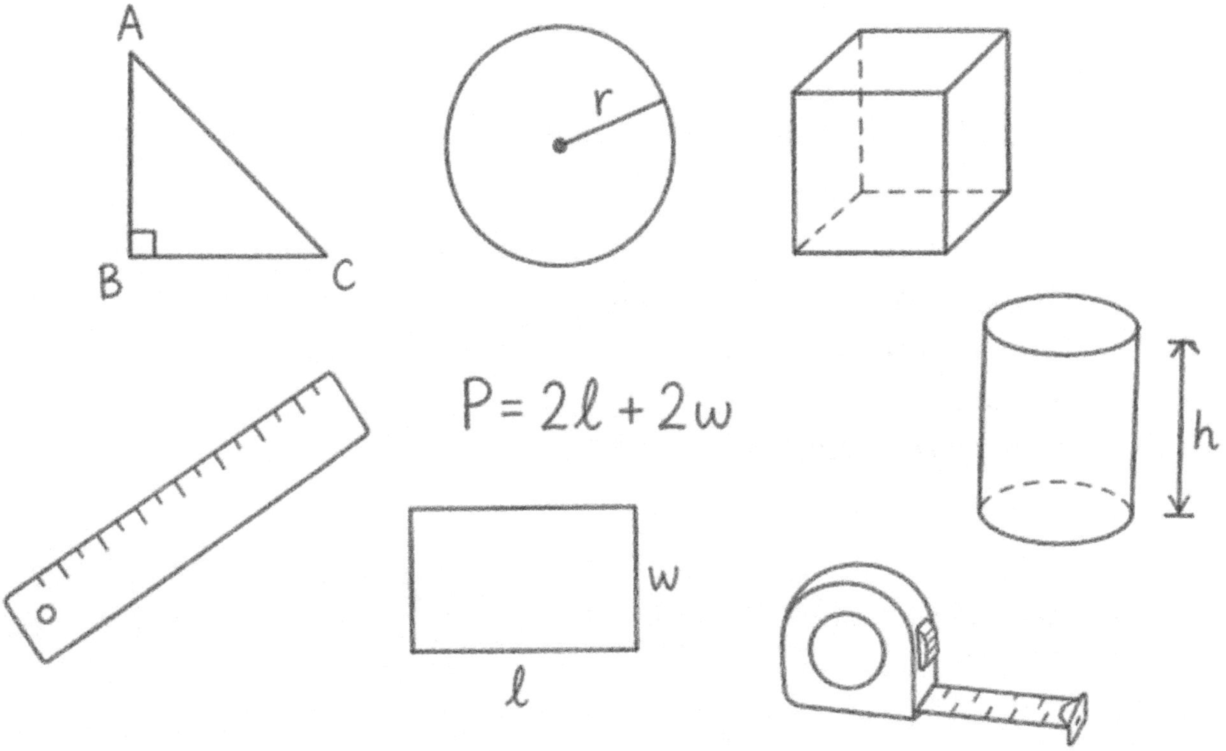

Triangle Area on Grid

QUESTION

A triangular sail is drawn inside a rectangle on a grid.

Each small square measures **3 m × 3 m**.

What is the area of the triangle?

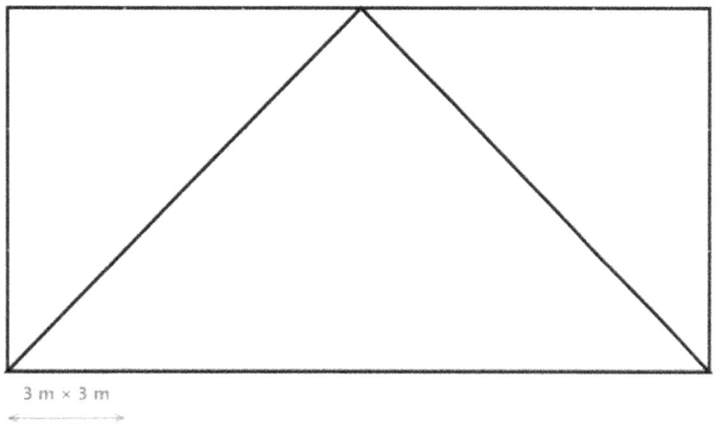

LET'S UNDERSTAND THE QUESTION TOGETHER

This question is about finding the **area of a shape using a grid**.

The triangle **does not** perfectly **fill each square**.
Some squares are fully inside the triangle, while others are only partly covered.

So, we need to carefully count:

- ✓ The **whole squares.**
- ✓ The **half-squares** (cut by the sloping sides).

We are trying to find the **total area inside the triangle**.

WHAT WE KNOW

- ✓ Each small square is **3 m × 3 m**
- ✓ Area of one square:
    - $3 \times 3 = 9 \text{ m}^2$
- ✓ From the diagram:
    - **6 whole squares**
    - **6 half-squares**

WHAT WE NEED TO FIND

The **area of the triangle**.

Our Plan

We will:

1. Count the **whole squares**
2. Count the **half-squares**
3. Divide the half-squares by 2 to convert them into square units
4. Add everything together
5. Multiply by the area of one square

This method works even if the number of half-squares is odd or even.

Step-by-Step Solution

Step 1: Area of one small square

$$3 \times 3 = 9 \text{ m}^2$$

Step 2: Count square units

From the diagram:

✓ Whole squares = **6**

✓ Half-squares = **6**

Convert half-squares:

$6 \div 2 = 3$

So total square units:

$6 + 3 = 9$

Step 3: Find the total area

$9 \times 9 = 81 \text{ m}^2$

Quick Check

Whole squares:

$6 \times 9 = 54 \text{ m}^2$

Half-squares:

$6 \div 2 = 3 \Rightarrow 3 \times 9 = 27 \text{ m}^2$

Total:

$54 + 27 = 81 \text{ m}^2$

Final Answer

The area of the triangle is 81 m^2, so the correct answer is B.

Common Mistake

Counting each half-square as a full square instead of dividing by 2.

Exam Tip

Always use this rule:

$$\textit{Total squares} = \textit{whole squares} + \frac{\textit{half-squares}}{2}$$

*This method works for **any shape on a grid**, even when the halves don't pair evenly.*

Addition of Capacity (L and mL)

QUESTION

At the school science fair, three robot cleaners collected spilled rainbow slime into separate tanks. At the end of the day, **all the slime is poured into one giant super-tank**. How much slime is in the super-tank altogether, in litres (L) and millilitres (mL)?

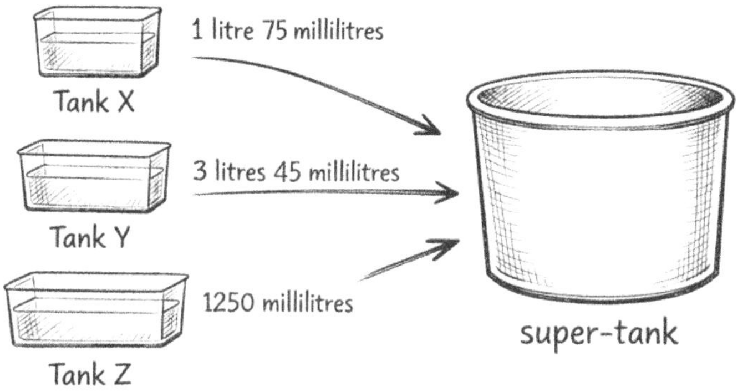

LET'S UNDERSTAND THE QUESTION TOGETHER

This question is asking us to find the **total amount of slime** in all three tanks.

Each tank has some slime in it.
We need to combine all of those amounts to find how much there is altogether.

So really, this is an **addition** question about liquid measurements.

WHAT WE KNOW

- ✓ Tank X = **1 L 75 mL**
- ✓ Tank Y = **3 L 45 mL**
- ✓ Tank Z = **1250 mL**
- ✓ **1 L = 1000 mL**

WHAT WE NEED TO FIND

We need to find the **total volume of slime** in the super-tank, written in **litres and millilitres**.

Our Plan

To make adding easier, we can change everything into **millilitres** first.
Then, once we find the total, we can change it back into **litres and millilitres**.

This works because all the measurements will be in the same unit.

Step-by-Step Solution

Step 1: Convert each amount to millilitres

Tank X:

- ✓ **1 L = 1000 mL**
- ✓ So, **1 L 75 mL = 1075 mL**

Tank Y:

- ✓ **3 L = 3000 mL**
- ✓ So, **3 L 45 mL = 3045 mL**

Tank Z:

- ✓ Already given as **1250 mL**

Step 2: Add the three amounts

$$1075 + 3045 + 1250 = 5370$$

So, the total is: 5370 mL

Step 3: Convert 5370 mL back into litres and millilitres

Since:

1000 mL = 1 L
5370 mL = 5 L 370 mL

Quick Check

Let's estimate:

- ✓ 1 L 75 mL is about **1 L**
- ✓ 3 L 45 mL is about **3 L**
- ✓ 1250 mL is about **1 L 250 mL**

So, the total should be a bit more than **5 L**.

Our answer is **5 L 370 mL**, which makes sense.

Final Answer

The super-tank contains **5 litres 370 millilitres** of slime altogether.

Alternative Method

Add litres and millilitres separately:

- ✓ Litres: $1 + 3 = 4$
- ✓ Millilitres: $75 + 45 + 1250 = 1370$**mL**

Convert: 1370mL = 1**L** 370**mL**

Total: $4 + 1 = 5$**L** 370**mL**

📌 Common Mistake

*Thinking **1250 mL** is **1 L 25 mL** instead of **1 L 250 mL.***

💡 Exam Tip

When units are mixed, first make them the same before adding.

Volume of a Rectangular Prism

QUESTION

A fish tank is shaped like a rectangular prism.

How much water does the larger fish tank hold?

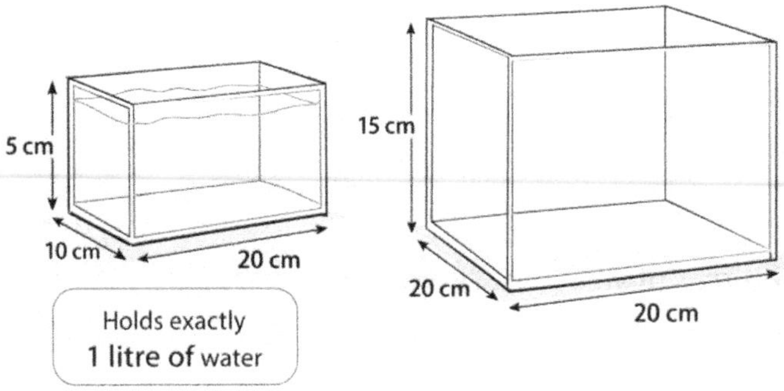

LET'S UNDERSTAND THE QUESTION TOGETHER

This question is about **capacity**, which means how much liquid something can hold.

We are told that the smaller tank holds **1 litre**.
That small tank is our guide.

Now we look at the larger tank and work out **how much water it can hold**.

So, the question is really asking:

WHAT WE KNOW

The smaller tank measures **5 cm × 10 cm × 20 cm**.

The smaller tank holds **1 litre**.

The larger tank measures **15 cm × 20 cm × 20 cm**.

Both tanks are rectangular boxes.

WHAT WE NEED TO FIND

We need to find **how many litres of water the larger tank can hold**.

Our Plan

We will find the **volume** of each tank using:

length × width × height

Then we will compare the larger tank to the smaller tank.

Since the smaller tank holds **1 litre**, we can use it as a unit to work out the larger tank's capacity.

Step-by-Step Solution

Step 1: Find the volume of the smaller tank

$5 \times 10 \times 20 = 1000$

So, the smaller tank has a volume of: $\mathbf{1000\ cm^3}$

We are told that this volume holds **1 litre**.

So: $\mathbf{1000\ cm^3 = 1\ litre}$

Step 2: Find the volume of the larger tank

$15 \times 20 \times 20 = 6000$

So, the larger tank has a volume of: $\mathbf{6000\ cm^3}$

Step 3: Work out how many litres this is

We know: $1000\ \text{cm}^3 = 1$ litre

So: $6000 \div 1000 = 6$

That means the larger tank holds: 6 litres

Quick Check

We found that the larger tank holds **6 litres**.

Now check by converting litres back to cubic centimetres:

$$6 \text{ litres} = 6 \times 1000 = 6000 \text{ cm}^3$$

Now find the volume of the larger tank:

$$15 \times 20 \times 20 = 6000 \text{ cm}^3$$

Both answers match.

So, **6 litres** is correct.

Final Answer

The larger tank can hold **6 litres of water**.

Common Mistake

Only multiplying two dimensions instead of all three. For volume, we must use ***length × width × height.***

Exam Tip

Remember this very useful fact:

$$1000\ cm^3 = 1\ litre$$

It can help you change volume into litres quickly.

Unit Conversion (Capacity)

QUESTION

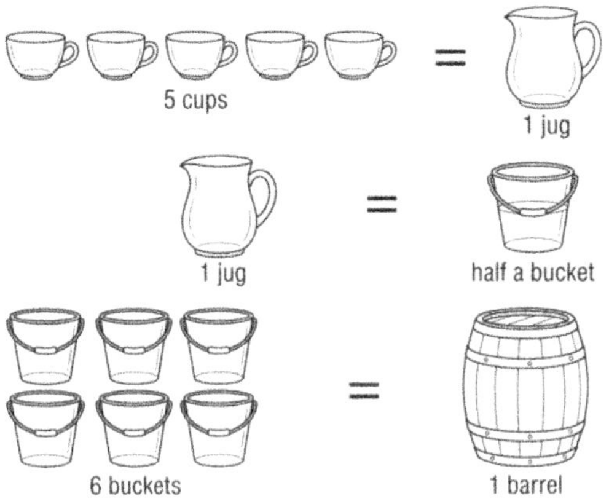

How many **cups** are equal to **1 barrel**?

LET'S UNDERSTAND THE QUESTION TOGETHER

This question is about changing from one unit to another.

We convert step by step from cups to jugs, buckets, and finally a barrel.

So, we need to work step by step through the unit changes until we can see how many cups make 1 barrel.

WHAT WE KNOW

- ✓ 5 cups = 1 jug.
- ✓ 1 jug = half a bucket.
- ✓ That means 2 jugs = 1 bucket.
- ✓ 6 buckets = 1 barrel.

WHAT WE NEED TO FIND

We need to find how many **cups** are equal to **1 barrel**.

Our Plan

We will convert one unit at a time.

First, change **buckets** into **jugs**.
Then change **jugs** into **cups**.

This works because each unit is connected to the next one.

Step-by-Step Solution

Step 1: Write the bucket fact in a more useful way.

Since **1 jug = half a bucket**, that means:

2 jugs = 1 bucket

Step 2: Change 6 buckets into jugs.

We know:

6 buckets = 1 barrel

And each bucket is **2 jugs**, so:

6 buckets = 6 × 2 = 12 jugs

Step 3: Change 12 jugs into cups.

We know:

1 jug = 5 cups

So:

12 jugs = 12 × 5 = 60 cups

Step 4: Write the final result clearly.

So: **1 barrel = 60 cups**

Quick Check

Let's check:

- ✓ 60 cups ÷ 5 = 12 jugs
- ✓ 12 jugs ÷ 2 = 6 buckets
- ✓ 6 buckets = 1 barrel

Everything matches, so our answer makes sense.

Final Answer

1 barrel is equal to **60 cups**.

📌 Common Mistake

*A common mistake is forgetting that if 1 jug is **half** a bucket, then **2 jugs** make 1 bucket.*

💡 Exam Tip

When unit conversions are chained together, rewrite each fact into the direction you need before starting. It makes the steps much easier to follow.

Dividing Length Measurements

QUESTION

A ribbon factory packs strips of ribbon into a bundle.
The diagram shows the **total thickness** of the bundle and **the number of strips**.

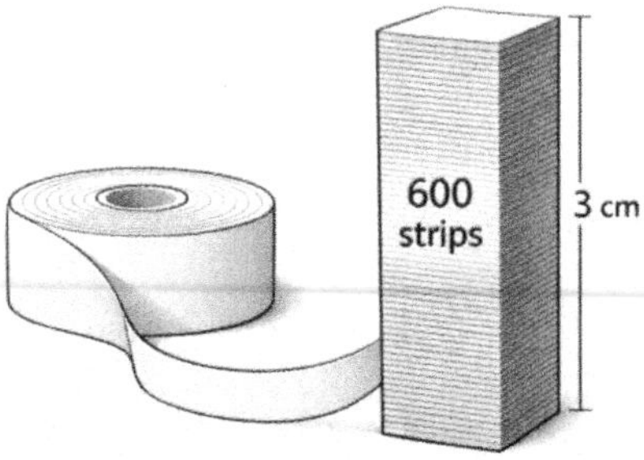

If **every ribbon strip** has the **same thickness**, what is the thickness of 1 strip?

LET'S UNDERSTAND THE QUESTION TOGETHER

This question is about a bundle of ribbon strips.

We know the whole bundle has a total thickness of 3 centimetres.

All 600 strips are the same thickness, so **each strip** must be only **a very small part of the whole bundle**.

We are trying to work out how thick just one strip is, in millimetres.

WHAT WE KNOW

600 ribbon strips together are **3 cm** thick.

Every strip has the **same thickness**.

We need to find the thickness of **1 strip**.

The answer must be in **millimetres.**

WHAT WE NEED TO FIND

We need to find the thickness of **one ribbon strip** in **millimetres**.

Our Plan

We will use **division**.

Since all the ribbon strips are the same thickness, we can:

- ✓ Divide the total thickness by the number of strips
- ✓ Then change the answer from centimetres to millimetres

This will tell us the thickness of one strip.

Step-by-Step Solution

Step 1: Convert 3 centimetres to millimetres

We know:

1 cm = 10 mm

So: 3 cm = 30 mm

Now we know that 600 strips together are **30 mm** thick.

Step 2: Divide the total thickness by 600

$30 \div 600 = 0.05$

So, the thickness of one strip is: 0.05 mm

Quick Check

Let's check by multiplying:

$600 \times 0.05 = 30$

So, 600 strips would be **30 mm** thick.

And: 30 mm = 3 cm

That matches the question, so the answer makes sense.

Final Answer

The thickness of one ribbon strip is **0.05 millimetres**.

📌 Common Mistake

Forgetting to change centimetres into millimetres before dividing.

💡 Exam Tip

If the final answer must be in millimetres, it is often quicker to ***convert first, then divide****.*

Measuring Angles with a Protractor

QUESTION

Sharalanda places one side of a stick so it matches a mark on a protractor. Another side of the stick lines up with a different mark.

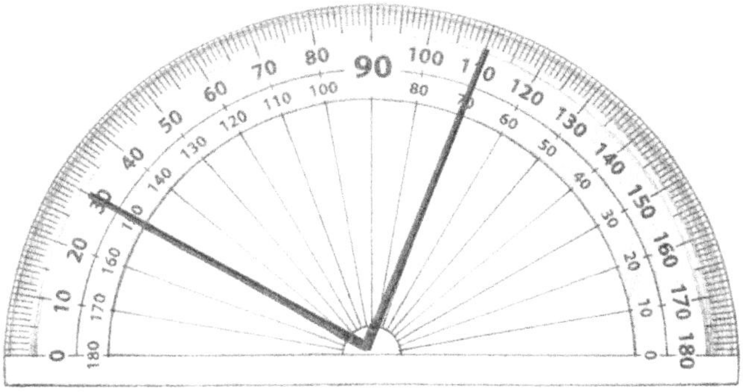

The angle between the two sides is **not obtuse**. What is its size?

LET'S UNDERSTAND THE QUESTION TOGETHER

We are looking at **two sides** of the stick on a protractor.

One side points to **30°.**

The other side points to **110°.**

The question tells us the angle is **not obtuse**. That means we are **not looking** for an angle bigger than **90°.**

So, we need to find the angle between the two markings and make sure it matches that clue.

WHAT WE KNOW

- ✓ One side is at **30°**
- ✓ The other side is at **110°**
- ✓ The angle is **not obtuse**

WHAT WE NEED TO FIND

We need to find the size of the angle between **30°** and **110°**.

Our Plan

We will subtract the smaller angle from the larger angle. Then we will check that the answer is **not obtuse**.

Step-by-Step Solution

Step 1: Identify the larger angle.

The larger angle is **110°**.

Step 2: Identify the smaller angle.

The smaller angle is **30°**.

Step 3: Subtract to find the angle between them.

$$110^\circ - 30^\circ = 80^\circ$$

Step 4: Check the clue.

An obtuse angle is **more than 90°**.
Our answer is **80°**, so it is **not obtuse**.

Quick Check

- ✓ From **30°** to **100°** is **70°**
- ✓ From **100°** to **110°** is **10°**
- ✓ $70^\circ + 10^\circ = 80^\circ$

The clue also fits because **80°** is less than **90°**.

Final Answer

The size of the angle between the two sides is **80°**.

Common Mistake

A common mistake is adding the two markings instead of finding the difference.
To find the angle between two sides on a protractor, subtract the smaller number from the larger number.

Exam Tip

When two sides point to different protractor marks, look for the bigger number first, then subtract the smaller one.

Angle Turns (Compass Directions)

QUESTION

A weathervane turns from **north-east** to **south** in the **shorter direction**.

Is the turn:

- ✓ Less than a quarter turn
- ✓ Between a quarter turn and a half turn
- ✓ Between a half turn and a three-quarter turn
- ✓ More than a three-quarter turn

Write the correct description.

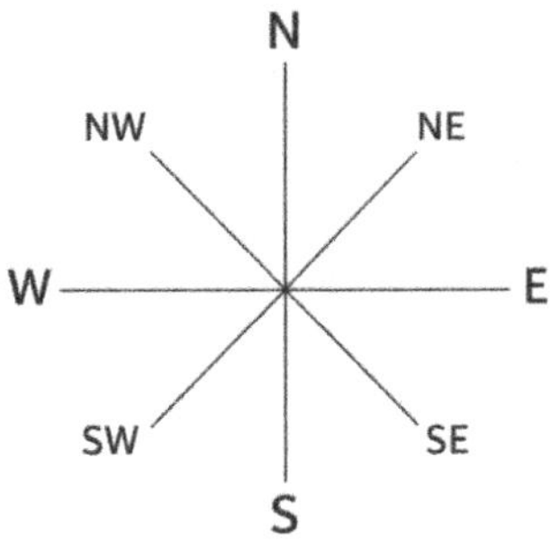

LET'S UNDERSTAND THE QUESTION TOGETHER

We are looking at a **turn**, not a length.

- ✓ A **quarter turn** is a **90° turn**.
- ✓ A **half turn** is a **180° turn**, so you face the opposite direction.
- ✓ A **three-quarter turn** is a **270° turn**.

The weathervane starts at **north-east** and ends at **south**.
It must turn in the **shorter direction**.
We need to decide how big that turn is and which description matches it.

WHAT WE KNOW

- ✓ The turn starts at **north-east**.
- ✓ The turn finishes at **south**.
- ✓ We use the **shorter direction**.
- ✓ A **quarter turn** is **90°**.
- ✓ A **half turn** is **180°**.
- ✓ A **three-quarter turn** is **270°**.

WHAT WE NEED TO FIND

We need to find which turn description matches the movement from **north-east** to **south**.

Our Plan

Work out the size of the turn from **north-east** to **south**, then compare it with **90°**, **180°**, and **270°**.

Step-by-Step Solution

Step 1: Identify where **north-east** is.

North-east is halfway between **north** and **east**.

Step 2: Work out the turn from **north-east** to **east**.

That is a turn of **45°**.

Step 3: Work out the turn from **east** to **south**.

That is a **quarter turn**, which is **90°**.

Step 4: Add the two parts.

$$45^\circ + 90^\circ = 135^\circ$$

So, the total turn is **135°**.

Step 5: Compare **135°** with the turn sizes.

- ✓ Quarter turn = **90°**
- ✓ Half turn = **180°**
- ✓ Three-quarter turn = **270°**

Step 6: Decide where **135°** fits.

135° is more than **90°** but less than **180°**.

So, the turn is **between a quarter turn and a half turn**.

Quick Check

A quarter turn is **90°**.
A half turn is **180°**.
Our turn is **135°**, which sits between them.

So, the answer makes sense.

Final Answer

The turn is **between a quarter turn and a half turn**.

Common Mistake

*A common mistake is taking the **longer way around** instead of the **shorter direction**.*
Always check which turn is smaller before you answer.

Exam Tip

For turn questions, remember these key angles:

- ✓ *Quarter turn = 90°*
- ✓ *Half turn = 180°*
- ✓ *Three-quarter turn = 270°*

Then compare your turn to these benchmark angles.

Turning Angles (Compass Directions)

QUESTION

Mika is standing in a playground, facing **south**.

She then turns **anticlockwise** to face a flagpole. The flagpole is **40° east of north**.

What is the angle of Mika's turn?

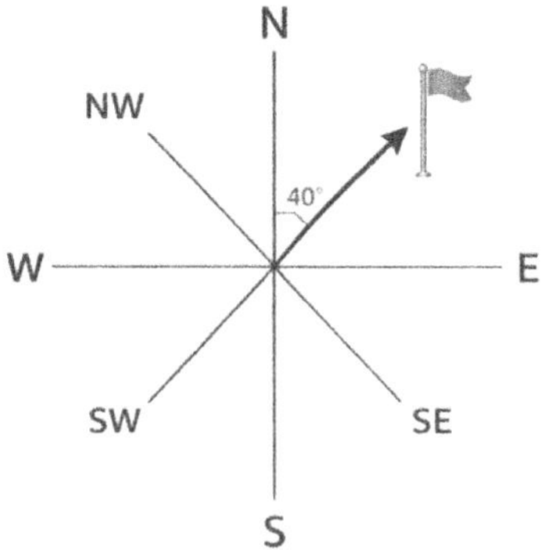

LET'S UNDERSTAND THE QUESTION TOGETHER

Mika starts by facing **south**.

Then she turns **anticlockwise**.

The flagpole is **40° east of north**. The flagpole is between **north** and **east**, and it is **40° away from north.**

It is **not exactly NE, but it is close to NE.**

We need to find the **angle of Mika's turn.**

WHAT WE KNOW

- ✓ Mika starts facing **south**
- ✓ She turns **anticlockwise**
- ✓ The flagpole is **40° east of north**

WHAT WE NEED TO FIND

We need to find the **angle Mika turns**.

Our Plan

We will:

1. Work out where the flagpole direction is
2. Follow the turn from south in the correct direction
3. Calculate the angle turned

Step-by-Step Solution

Step 1: Find the turn from **south** to **east**.

Turning anticlockwise from south to east is: 90°

Step 2: Find the turn from **east** to the flagpole.

From east to north is: 90°

The flagpole is **40° away from north** toward east.

So, the turn from east to the flagpole is:

$90^\circ - 40^\circ = 50^\circ$

Step 3: Add the two parts of the turn.

$90^\circ + 50^\circ = 140^\circ$

So, Mika turns: 140°

Quick Check

From south to north is: 180°

The flagpole is **40° away from north** toward east, so it is **40° less than that half-turn**.

$180^\circ - 40^\circ = 140^\circ$

That matches our answer.

Final Answer

Mika's turn is **140° anticlockwise**.

Common Mistake

*A common mistake is thinking **east of north** means exactly **NE**.*
It does not. You must use the exact angle given in the question.

Exam Tip

*When you see directions like **east of north**, start at the first direction named, then move toward the second direction by the angle given.*

Final Direction After Turns

QUESTION

A wind-up toy starts by facing in the direction shown in the diagram.

It then turns 37 quarter-turns clockwise, followed by 18 quarter-turns anti-clockwise.

Which direction is the toy facing now?

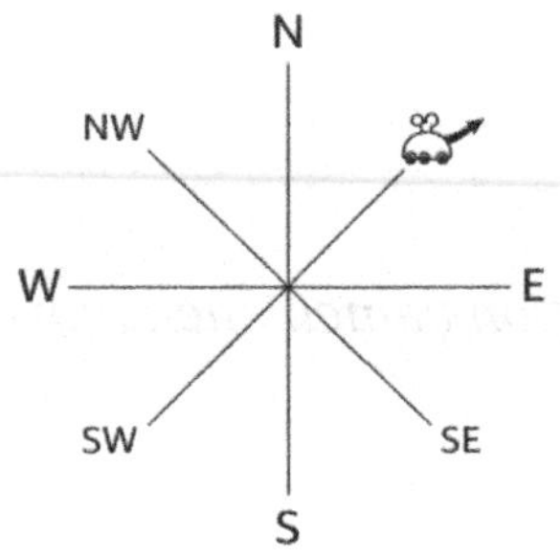

LET'S UNDERSTAND THE QUESTION TOGETHER

The toy starts by facing **north-east**.

A **quarter-turn** means a turn of **90°**.

The toy first turns **clockwise**, then it turns **anti-clockwise**.

We need to work out the **final direction** after all the turns are finished.

WHAT WE KNOW

The toy starts facing **north-east.**

It turns **37 quarter-turns clockwise**.

Then it turns **18 quarter-turns anti-clockwise.**

4 quarter-turns make one full turn.

WHAT WE NEED TO FIND

We need to find the **direction the toy is facing at the end**.

Our Plan

We will find the **net turn** first.

To help with this:

- ✓ **Clockwise = positive**
- ✓ **Anti-clockwise = negative**

Then we will simplify the number of quarter-turns and count the final direction.

Step-by-Step Solution

Step 1: Find the net rotation

The toy turns:

$+37 + (-18)$
$37 - 18 = 19$

So, the toy turns **19 quarter-turns clockwise overall**.

Step 2: Simplify using full turns

A full turn is **4 quarter-turns**.

$19 \div 4 = 4$ remainder 3

This means the 19 quarter-turns can be split into:

- **16 quarter-turns** = 4 full turns
- **3 extra quarter-turns**

After **16 quarter-turns**, the toy is back to its **starting direction**, because every 4 quarter-turns make one full turn.

So only the **3 extra quarter-turns clockwise** matter.

Step 3: Count 3 quarter-turns clockwise from north-east

Starting at **north-east**:

- **1 quarter-turn clockwise**: south-east
- **2 quarter-turns clockwise**: south-west
- **3 quarter-turns clockwise**: north-west
- **4 quarter-turns clockwise**: back to north-east

So, after **3 quarter-turns clockwise**, the toy is facing **north-west**.

Quick Check

We said the toy finishes at **north-west**.

To check, reverse the last **3 quarter-turns clockwise** by turning **3 quarter-turns anti-clockwise**:

north-west → south-west → south-east → north-east

This takes us back to the starting direction, **north-east**, so the answer is correct.

Final Answer

The toy is facing **north-west**.

📌 Common Mistake

A common mistake is forgetting that:

- ***Clockwise = positive***
- ***Anti-clockwise = negative***

*It is also easy to forget that **4 quarter-turns** bring the toy back to the **starting direction**.*

💡 Exam Tip

*When a question has turns in both directions, first find the **net turn**.*

*Then remember that **every 4 quarter-turns make a full turn**, so the toy comes back to the same direction after each full cycle.*

Clock Hand Angle Turns

QUESTION

What angle does the **hour hand** turn **clockwise** from **4:30 pm** to **5:00 pm**?

LET'S UNDERSTAND THE QUESTION TOGETHER

We are following the **hour hand only**.

The hand starts at **4:30 pm** and ends at **5:00 pm**.

We are not finding the angle between the two hands.

We are finding how far the **hour hand turns clockwise** during that time.

WHAT WE KNOW

The time starts at **4:30 pm**.

The time ends at **5:00 pm**.

We are looking at the **hour hand**.

We need to find how far it turns **clockwise.**

WHAT WE NEED TO FIND

The angle the **hour hand** turns clockwise from **4:30 pm** to **5:00 pm**.

Our Plan

First, we will use the fact that the **hour hand moves a fixed amount in 1 hour**.

Then, we will work out what fraction of an hour passes from **4:30 pm to 5:00 pm**.

Finally, we will use that fraction to find the number of **degrees** the hour hand turns.

Step-by-Step Solution

Step 1: Find how far the hour hand moves in 1 hour.

The hour hand moves from one number to the next in 1 hour.

A clock is divided into 12 equal parts, so one hour step is:

$$360^\circ \div 12 = 30^\circ$$

So, the hour hand moves **30° in 1 hour**.

Step 2: Find the time from 4:30 pm to 5:00 pm.

From **4:30 pm to 5:00 pm** is: 30 minutes

That is **half of 1 hour**.

Step 3: Find half of 30°.

If the hour hand moves **30° in 1 hour**, then in **half an hour** it moves:

$$30^\circ \div 2 = 15^\circ$$

Step 4: State the answer.

So, the hour hand turns: 15°clockwise.

Quick Check

In 1 full hour, the hour hand moves **30°**.

Only **half an hour** passes here, so the turn should be **half of 30°**.

That matches our answer.

Final Answer

The hour hand turns **15° clockwise** from **4:30 pm** to **5:00 pm**.

📌 Common Mistake

*A common mistake is answering **30°**.*

*That is the angle the hour hand turns in **1 full hour**, not in **30 minutes**.*

*From **4:30 pm to 5:00 pm** is only **half an hour**, so the correct angle is **15°**.*

💡 Exam Tip

For clock questions, first decide whether you are finding:

- ✓ *The movement of **one hand**, or*
- ✓ *The angle **between two hands***

That helps you avoid solving the wrong thing.

Finding Rectangle Length

QUESTION

A festival organiser uses **6.2 m** of lighting wire to make a rectangular border around a display area.

What is the display **area's length** in metres?

LET'S UNDERSTAND THE QUESTION TOGETHER

The **6.2 m** of lighting wire goes all the way **around** the display area.

So, **6.2 m** is the rectangle's **perimeter**.

The width is **110 cm**.

We need to find the missing **length**.

The question asks for the final answer in **metres**.

This means we need to pay close attention to the **units**.

WHAT WE KNOW

- ✓ The perimeter is **6.2 m**
- ✓ The width is **110 cm**
- ✓ A rectangle has **2 lengths** and **2 widths**
- ✓ The final answer must be in **metres**

WHAT WE NEED TO FIND

The **length of the display area in metres**.

Our Plan

First, convert the perimeter to **centimetres**.

Then use the perimeter of the rectangle to find the missing length.

Finally, convert the answer back to **metres**.

Step-by-Step Solution

Step 1: Convert the perimeter to centimetres.

$6.2 \text{ m} = 620 \text{ cm}$

So, the perimeter is **620 cm**.

Step 2: Write the perimeter rule for a rectangle.

$2(L + W) = 620$

Step 3: Substitute the width.

Since the width is 110 cm,

$2(L + 110) = 620$

Step 4: Divide both sides by 2.

$L + 110 = 310$

Step 5: Subtract 110 from both sides.

$L = 310 - 110$
$L = 200$

So, the length is **200 cm**.

Step 6: Convert the length to metres.

$200 \text{ cm} = 2 \text{ m}$

So, the length is **2 m**.

Quick Check

Check in centimetres:

$$2(200 + 110) = 2(310) = 620$$

That matches the perimeter.

And $200 \text{ cm} = 2 \text{ m}$, so the answer in metres is correct.

Final Answer

The length of the display area is **2 m**.

Common Mistake

*A common mistake is finding **200 cm** and forgetting to convert it to **metres** at the end.*

Exam Tip

*Always check the **unit** asked for in the final answer before you finish.*

Total Edge Length (Rectangular Prism)

QUESTION

A storage box is shaped like a **rectangular prism**. Its length is **2 cm**, its width is **5 cm**, and its height is **7 cm**.

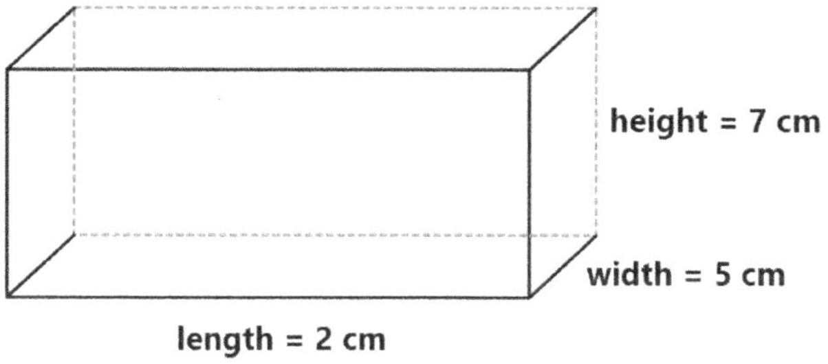

What is the total length of all the edges of the storage box?

LET'S UNDERSTAND THE QUESTION TOGETHER

A rectangular prism has edges in **three directions**: length, width, and height.

The box has:

- ✓ A **length** of **2 cm**
- ✓ A **width** of **5 cm**
- ✓ A **height** of **7 cm**

We need to find the **sum of all the edges** on the box.

WHAT WE KNOW

- ✓ Length – **2 cm**.
- ✓ Width = **5 cm**.
- ✓ Height = **7 cm**.
- ✓ A **rectangular prism** has **4 edges** of **each dimension**:

    1. 4 edges are the length
    2. 4 edges are the width
    3. 4 edges are the height

WHAT WE NEED TO FIND

We need to find the **total length of all 12 edges.**

Our Plan

We will use:

$4 \times$ (length + width + height)

This works because each of the three dimensions appears on **4 edges.**

Step-by-Step Solution

Step 1: Write the formula.

$4 \times (\text{length} + \text{width} + \text{height})$

We use this because a rectangular prism has **4 edges of each dimension.**

Step 2: Add inside the brackets, then multiply.

$4 \times (2 + 5 + 7) = 4 \times 14 = 56$

So, the total length of all the edges is **56 cm.**

Quick Check

We can check by finding the total for each dimension:

- ✓ $4 \times 2 = 8$
- ✓ $4 \times 5 = 20$
- ✓ $4 \times 7 = 28$

$8 + 20 + 28 = 56$

So, our answer is correct.

Final Answer

The total length of all the edges of the storage box is **56 cm**.

Common Mistake

A common mistake is to add the three dimensions only once:

$$2 + 5 + 7 = 14$$

*But **14 cm** is not the total length of all the edges.*
*Each dimension appears on **4 edges**, so we must multiply by **4**.*

Exam Tip

For any rectangular prism, the total length of all edges is:

$$4 \times (length + width + height)$$

This is a quick and useful shortcut in tests.

Perimeter of Smaller Squares

QUESTION

A square garden has a perimeter of **40 m**.
It is divided into **4 equal smaller squares** as shown in the drawing.
What is the perimeter of **1 smaller square**?

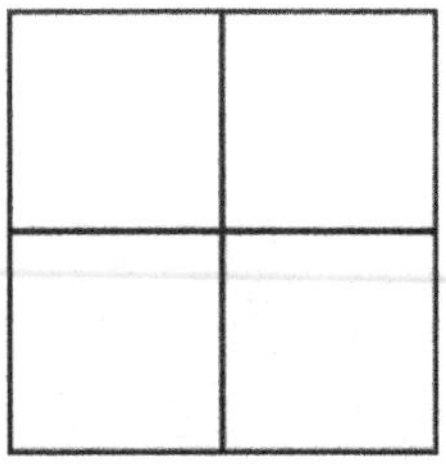

LET'S UNDERSTAND THE QUESTION TOGETHER

We are starting with **one big square**.

Its **perimeter** is 40 m.

Then it is split into **4 equal smaller squares**, so each smaller square must have the same side length.

We are not finding the area. We are finding the **perimeter of one small square**, which means the distance all the way around just **one** of the small pieces.

WHAT WE KNOW

- ✓ The big square has a perimeter of **40 m.**
- ✓ A square has **4 equal sides.**
- ✓ The big square is divided into **4 equal smaller squares.**
- ✓ Each smaller square will have side lengths that are **half** of the big square's side length.

WHAT WE NEED TO FIND

The perimeter of **one smaller square.**

Our Plan

- ✓ First, find the side length of the big square.
- ✓ Then work out the side length of one smaller square.
- ✓ Finally, find the perimeter of that smaller square.

Step-by-Step Solution

Step 1: Find the side length of the big square.

A square has 4 equal sides.

$$40 \div 4 = 10$$

So, the side length of the big square is **10 m**.

Step 2: Find the side length of one smaller square.

The big square is split into **2 smaller squares across** and **2 smaller squares down**.

Each smaller square has a side length equal to half the side length of the big square.

$$10 \div 2 = 5$$

So, the side length of one smaller square is **5 m**.

Step 3: Find the perimeter of one smaller square.

A square's perimeter is 4 times its side length.

$4 \times 5 = 20$

So, the perimeter of one smaller square is **20 m**.

Quick Check

A small square has side length **5 m**.

Its perimeter is:

$$5 + 5 + 5 + 5 = 20$$

So, our answer checks out.

Final Answer

The perimeter of one smaller square is **20 m**.

Common Mistake

Some students find the side length of the big square and stop too early.

Others divide the big perimeter by 4 and think that answer is the perimeter of a small square.

*Be careful to work out the **small square's side length first**, then find its perimeter.*

Exam Tip

*When a square is divided into **4 equal smaller squares**, think of it as **2 across and 2 down**. That helps you see why each small side is **half** the big side.*

Perimeter Change After Rearrangement

QUESTION

Nai has a square piece of cloth with side length **30 cm**.
She cuts the cloth into **2 equal rectangular pieces** and rearranges them to make **1 rectangle**.

Find the difference in perimeter between the original square and the new rectangle.

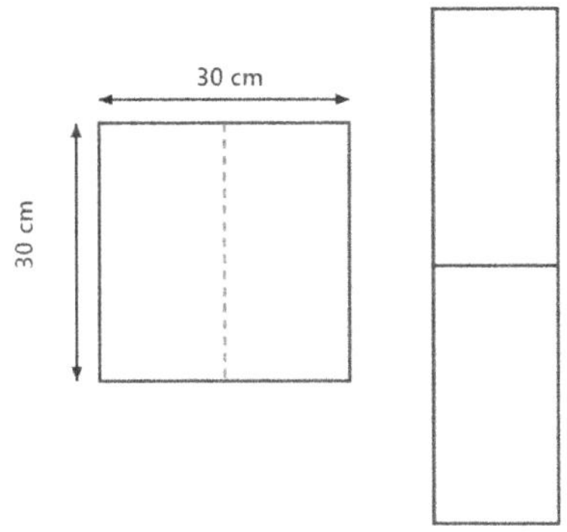

LET'S UNDERSTAND THE QUESTION TOGETHER

The cloth starts as **one square**.

Each side of the square is **30 cm**.

It is cut into **2 equal rectangular pieces**, so the two pieces are the same size.

After that, the 2 pieces are rearranged to make **one rectangle**.

We are comparing the **perimeter of the original square** and the **perimeter of the new rectangle**.

The question is asking how much the perimeter changes.

WHAT WE KNOW

- ✓ The original cloth is a **square**.
- ✓ The side length of the square is **30 cm**.
- ✓ The cloth is cut into **2 equal rectangular pieces**.
- ✓ Each rectangular piece has:
  - **length = 30 cm**
  - **width = 15 cm**
- ✓ The 2 pieces are rearranged to make **1 rectangle**.
- ✓ We need the **difference in perimeter.**

WHAT WE NEED TO FIND

We need to find how many centimetres the perimeter changes by after the square cloth is cut and rearranged.

Our Plan

First, find the perimeter of the **square** by multiplying the side length by **4**.

Next, work out the **length** and **width** of the new rectangle.

Then, find the perimeter of the **rectangle** using **2 × (length + width)**.

Finally, subtract the 2 perimeters to find the difference.

Step-by-Step Solution

Step 1: Find the perimeter of the original square.

A square has 4 equal sides.

Perimeter of square
$= 4 \times 30 =$ **120 cm**

So, the perimeter of the original square is **120 cm**.

Step 2: Work out the dimensions of each rectangular piece.

The square is **30 cm by 30 cm**.

It is cut into 2 equal rectangles.

One side stays the same, so the **length = 30 cm**.

The other side is halved:

Width = $30 \div 2 =$ **15 cm**

Each piece is 30 cm by 15 cm.

Step 3: Work out the dimensions of the new rectangle.

The 2 rectangular pieces are rearranged to make 1 rectangle.

The 30 cm sides join end to end.

New length:

$= 30 + 30 =$ **60 cm**

The width stays the same.

New width = **15 cm**

So, the new rectangle is:

- **length = 60 cm**
- **width = 15 cm**

So, the new rectangle is **60 cm long and 15 cm wide**.

Step 4: Find the perimeter of the new rectangle.

Perimeter of rectangle:

$= 2 \times (\text{length} + \text{width})$
$= 2 \times (60 + 15)$
$= 2 \times 75$
= **150 cm** (So, the perimeter of the new rectangle is **150 cm**.)

Step 5: Find the difference in perimeter.

$= 150 - 120$
= **30 cm**

Quick Check

The new rectangle is longer and thinner than the square, so it makes sense that its perimeter is bigger.

Original square perimeter = **120 cm**
New rectangle perimeter = **150 cm**

Difference = **30 cm**. So, the answer is reasonable.

Final Answer

The difference in perimeter is **30 cm**.

Common Mistake

*A common mistake is halving **both** side lengths after cutting.*

*Only **one side** is halved.*

*So, each piece is **30 cm long and 15 cm wide**, not **15 cm by 15 cm**.*

Exam Tip

*For a **square**, all 4 sides are equal, so perimeter is easy to find.*

*For a **rectangle**, make sure you know the **length** and **width** before using the perimeter formula.*

Perimeter of Enlarged Rectangle

QUESTION

Nai is decorating a rectangular display board for a school project.

The board is **3 m long** and **2 m wide**.

A glittery ribbon border extends **400 mm beyond each side** of the board.

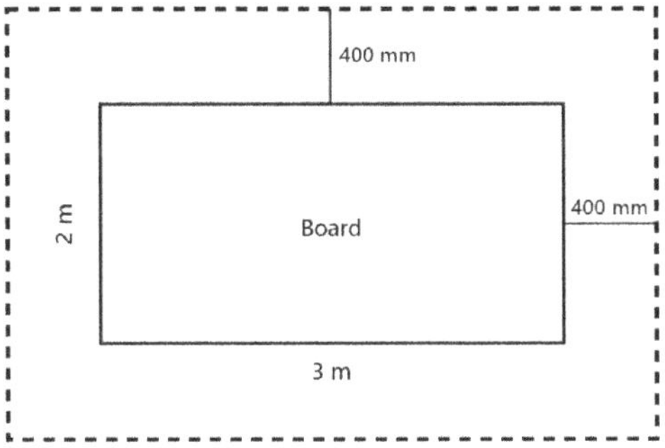

What is the distance around the **outside edge of the ribbon**?

LET'S UNDERSTAND THE QUESTION TOGETHER

The display board is a **rectangle**.

The ribbon goes **past the edge** of the board on **every side**.

That means the ribbon makes a **bigger rectangle** around the board.

We are not finding the distance around the board itself. We are finding the distance around the **outside edge of the ribbon**.

WHAT WE KNOW

- ✓ The board is **3 m long**.
- ✓ The board is **2 m wide**.
- ✓ The ribbon extends **400 mm on each side**.
- ✓ We need the distance around the **outside edge** of the ribbon.

WHAT WE NEED TO FIND

We need to find the **perimeter of the larger rectangle** made by the ribbon.

Our Plan

First, change all the measurements to the **same unit**.

Then work out the **new length** and **new width** of the ribbon border.

After that, find the **perimeter** of the larger rectangle.

Step-by-Step Solution

Step 1: Change the board measurements to centimetres.

It is easiest to work in **centimetres**.

$3 \text{ m} = 300 \text{ cm}$
$2 \text{ m} = 200 \text{ cm}$

So, the board is:

- **300 cm long**
- **200 cm wide**

Step 2: Change 400 mm to centimetres.

We also need to convert the extra ribbon measurement.

$400 \text{ mm} = 40 \text{ cm}$

So, the ribbon extends **40 cm on each side.**

Step 3: Work out the new length.

The ribbon extends **40 cm at one end** and **40 cm at the other end.**

So, we add **40 cm twice**:

$300 + 40 + 40 = 380$

The new length is **380 cm.**

Step 4: Work out the new width.

The ribbon extends **40 cm on one side** and **40 cm on the other side.**

So, we add **40 cm twice**: $200 + 40 + 40 = 280$

The new width is **280 cm.**

Step 5: Find the perimeter of the larger rectangle.

Now use the perimeter formula:

Perimeter: $2 \times (\text{length} + \text{width})$

$$= 2 \times (380 + 280)$$
$$= 2 \times 660$$
$$= 1320$$

So, the perimeter is **1320 cm.**

Step 6: Convert the answer to metres.

1320 cm = 13.2 m

Quick Check

The outside rectangle is **3.8 m by 2.8 m.**

Checking the perimeter:

$3.8 + 2.8 + 3.8 + 2.8 = 13.2$

So, the answer makes sense.

Final Answer

The distance around the outside edge of the ribbon is **13.2 metres.**

📌 Common Mistake

*A common mistake is adding **400 mm only once**.*

*But the ribbon extends on **both sides** of the length and **both sides** of the width, so the extra amount must be added **twice** each time.*

💡 Exam Tip

*When a border extends **beyond each side**, picture one extra piece on the **left** and one on the **right**, or one on the **top** and one on the **bottom**. That helps you remember why the extra amount is added **twice**.*

Tablecloth Perimeter (one-sided overhang)

QUESTION

Sharalanda has a rectangular picnic table that is **1.8 m long** and **90 cm wide**.

She buys a rectangular cloth to cover it. The cloth hangs down **15 cm** over both of the **shorter sides**, but it does **not** hang over the longer sides.

Sharalanda wants to sew ribbon all the way around the outside edge of the cloth.

How many centimetres of ribbon does she need?

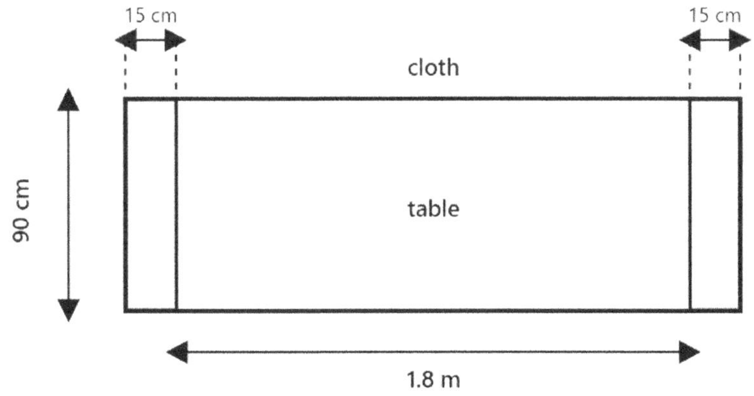

LET'S UNDERSTAND THE QUESTION TOGETHER

This question is about a **rectangular cloth** covering a **rectangular table**.

The cloth hangs over the two ends of the table by 15 cm each.

There is **no overhang** on the longer sides.

The question does not say the word **perimeter**, but when it asks for ribbon **all the way around the outside edge**, it is really asking for the **perimeter of the cloth**.

So, we need to work out the cloth's size first, then find the distance around it.

WHAT WE KNOW

- ✓ The table is **1.8 m long**.
- ✓ The table is **90 cm wide**.
- ✓ The cloth hangs down **15 cm** over each **shorter side**.
- ✓ The cloth does **not** hang over the longer sides.
- ✓ The ribbon goes **all the way around the outside edge**.
- ✓ The answer must be in **centimetres.**

WHAT WE NEED TO FIND

We need to find how many **centimetres of ribbon** are needed to go around the edge of the cloth.

Our Plan

First, work out the **full dimensions of the cloth.**

Then find the **perimeter of the cloth,** because "ribbon all the way around the outside edge" means we need the distance around the shape.

Step-by-Step Solution

Step 1: Change the length into centimetres.

$$1.8 \text{ m} = 180 \text{ cm}$$

So, the table measures:

- ✓ Length = **180 cm**
- ✓ Width = **90 cm**

Step 2: Add the overhang on the two shorter sides.

The cloth hangs down **15 cm** on one **shorter** side and **15 cm** on the other shorter side.

$$15 + 15 = 30 \text{ cm}$$

So, the cloth's new length is:

$180 + 30 = 210$ cm

Step 3: Work out the width of the cloth.

There is **no overhang** on the longer sides, so the width stays the same (90cm).

So, the cloth measures: **210 cm by 90 cm**

Step 4: Find the perimeter of the cloth.

The perimeter of a rectangle is:

$$2 \times (\text{length} + \text{width})$$

$$2 \times (210 + 90)$$
$$2 \times 300 = 600$$

So, the perimeter is: 600 cm

Quick Check

The cloth is about **210 cm by 90 cm**.

Going around it means:

- ✓ Two long sides: about $2 \times 210 = 420$
- ✓ Two short sides: about $2 \times 90 = 180$

$$420 + 180 = 600$$

That matches our answer.

Final Answer

Sharalanda needs **600 cm of ribbon**.

Common Mistake

*A common mistake is adding **15 cm only once**.*
*The cloth hangs over **both** shorter sides, so you must add **15 cm + 15 cm**.*

Exam Tip

*When a question mixes **metres** and **centimetres**, convert everything to **one unit first** before solving.*

Area Scale Drawing (Rectangle)

QUESTION

A school is making a rectangular vegetable garden.
On this sketch, **2 cm represents 6 m in real life**.
Use the sketch to find the real area of the garden in square metres.

LET'S UNDERSTAND THE QUESTION TOGETHER

This question is about a **garden sketch** and the **real garden**.

The measurements in centimetres are only the sizes **on the sketch**.

We need to use the **scale** to find the real length and real width.

After that, we work out the **area** of the real garden.

The answer must be in **square metres**.

WHAT WE KNOW

- ✓ The garden is a **rectangle**.
- ✓ Sketch length = **8 cm**.
- ✓ Sketch width = **5 cm**.
- ✓ **2 cm on the sketch = 6 m in real life.**

WHAT WE NEED TO FIND

We need to find the **real area** of the garden in $\mathbf{m^2}$.

Our Plan

First, convert the sketch measurements into real-life metres.

Then find the area of the rectangle using:

Area = length × width

Step-by-Step Solution

Step 1: Work out how many metres 1 cm represents

We know:

$$2 \text{ cm} = 6 \text{ m}$$

Now divide by 2:

$$1 \text{ cm} = 3 \text{ m}$$

Step 2: Find the real length

The sketch length is **8 cm**.

Since **1 cm = 3 m**:

$8 \text{ cm} = 8 \times 3 = 24 \text{ m}$

So, the real length is **24 m**.

Step 3: Find the real width

The sketch width is **5 cm**.

Since **1 cm = 3 m**:

$5 \text{ cm} = 5 \times 3 = 15 \text{ m}$

So, the real width is **15 m**.

Step 4: Find the real area

Area of a rectangle:

Area = length × width

So:

Area = 24 × 15

Area = 360

So, the real area is **360 m²**.

Quick Check

Let's check the answer.

A rectangle that is about **24 m by 15 m** should have an area around:

$25 \times 15 = 375$

Our answer of **360 m²** is close, so it makes sense.

Final Answer

The real area of the garden is **360 square metres.**

Common Mistake

A common mistake is finding the real length correctly but forgetting that area involves two dimensions. Both length and width must be scaled before calculating the area.

Exam Tip

*In scale questions, convert the **side lengths first**, then find the area. That helps avoid mistakes with square units.*

Scale Drawing Distances (Chessboard)

QUESTION

Mika is designing a giant outdoor chessboard.

On Mika's drawing, **each square cell is 1 cm by 1 cm**.
Also, **each 1 centimetre represents 3 metres in real life**.

Mika measures two straight distances on the drawing:

- ✓ From the **white king** to the **black rook** is **4 cm**
- ✓ From the **white queen** to the **black knight** is **2.5 cm**

What is the actual distance from the **white king** to the **black rook**?

What is the actual distance from the **white queen** to the **black knight**?

LET'S UNDERSTAND THE QUESTION TOGETHER

This is a **scale drawing** of a chessboard.

The drawing uses **square cells**, and each cell is **1 cm by 1 cm**.
That helps us understand the size of each square on the drawing.

We are also told the **scale**:
1 cm on the drawing stands for 3 m in real life.

So, the distances measured on the drawing are much smaller than the real distances.

We need to change each **drawing distance** into a **real-life distance**.

WHAT WE KNOW

- ✓ **Each square cell on the drawing is 1 cm by 1 cm**
- ✓ **1 cm on the drawing = 3 m in real life**
- ✓ The distance from the **white king to the black rook** is **4 cm**
- ✓ The distance from the **white queen to the black knight** is **2.5 cm**

WHAT WE NEED TO FIND

We need to find:

- ✓ The actual distance from the **white king** to the **black rook**
- ✓ The actual distance from the **white queen** to the **black knight**

Our Plan

We will use the scale.

Since **1 cm represents 3 m**, we multiply each distance on the drawing by **3** to find the real distance.

Step-by-Step Solution

Step 1

Find the actual distance from the **white king** to the **black rook**.

Since **1 cm = 3 m**:

$4 \times 3 = 12$

So, the actual distance from the **white king** to the **black rook** is **12 m**.

Step 2

Find the actual distance from the **white queen** to the **black knight**.

Since **1 cm = 3 m**:

$2.5 \times 3 = 7.5$

So, the actual distance from the **white queen** to the **black knight** is **7.5 m**.

Quick Check

If **1 cm = 3 m**, then:

- **4 cm** becomes **12 m**
- **2.5 cm** becomes **7.5 m**

Both answers match the scale correctly.

Final Answer

The actual distance from the **white king** to the **black rook** is **12 m**, and the actual distance from the **white queen** to the **black knight** is **7.5 m**.

Common Mistake

*A common mistake is mixing up the **cell size on the drawing** with the **real-life scale**.*

Remember:

- ✓ ***1 cm by 1 cm** tells us the size of each cell on the drawing*
- ✓ ***1 cm = 3 m** tells us how to convert to real life*

Exam Tip

*When a question gives both a **grid size** and a **scale**, ask yourself:*

"Which fact helps me measure, and which fact helps me convert?"

That helps you use the right information.

Composite Area (Rectangle and Triangle)

QUESTION

Inside the noticeboard, there is a piece of paper with a rectangle.
Inside the rectangle, there is a triangular sticker.
The part of the rectangle not covered by the sticker has a **grey area** of **114 cm^2**.

Using the information shown in the image, find the **area of the triangular sticker**.

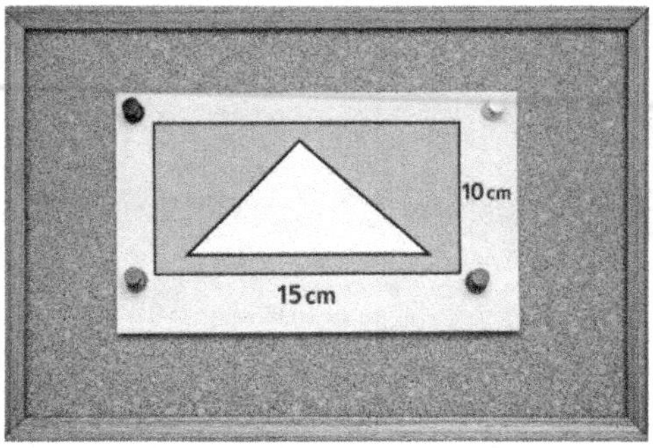

LET'S UNDERSTAND THE QUESTION TOGETHER

We have a **rectangle** and a **triangle inside it**.

The **uncovered part** means the part of the rectangle that is left over after the triangular sticker covers part of the rectangle.

So:

- ✓ The whole rectangle has one area
- ✓ The uncovered part has another area
- ✓ The triangle is the part we need to find

The question is asking for the **area of the triangle only**.

WHAT WE KNOW

- ✓ The rectangle is **15 cm** wide.
- ✓ The rectangle is **10 cm** high.
- ✓ The uncovered part of the rectangle is **114 cm^2**.

We do not need to use the triangle area formula because we are given the leftover area.

WHAT WE NEED TO FIND

The **area of the triangular sticker**.

Our Plan

1. First, find the area of the whole rectangle.
2. Then subtract the uncovered area.
3. That will give us the area of the triangle.

Step-by-Step Solution

Step 1: Find the area of the rectangle.

$$\text{Area of rectangle} = \text{length} \times \text{width}$$

$$15 \times 10 = 150$$

So, the area of the rectangle is **150 cm²**.

Step 2: Write down the uncovered area.

The uncovered area is **114 cm²**.

Step 3: Subtract to find the triangle's area.

$$150 - 114 = 36$$

So, the area of the triangle is **36 cm²**.

Quick Check

The whole rectangle is **150 cm²**.

If the triangle is **36 cm²**, then the uncovered part would be:

$$150 - 36 = 114$$

That matches the question, so our answer is correct.

Final Answer

The area of the triangular sticker is **36 cm²**.

Common Mistake

Some students subtract the numbers the wrong way around.

Remember to do ***whole rectangle area – uncovered area****.*

Exam Tip

When one shape is ***inside*** *another, think about whether you need to* ***add*** *or* ***subtract*** *areas. Here, the triangle is part of the rectangle, so we subtract the leftover area from the whole rectangle.*

Area of Triangle in Rectangle

QUESTION

Nai is designing a rectangular sports banner that is **16 m wide** and **8 m high**.

A diagonal ribbon across the banner divides the rectangle into **two triangles**. One triangle is **smaller** and one triangle is **larger**, as shown in the image.

What is the **area of the smaller triangular section**?

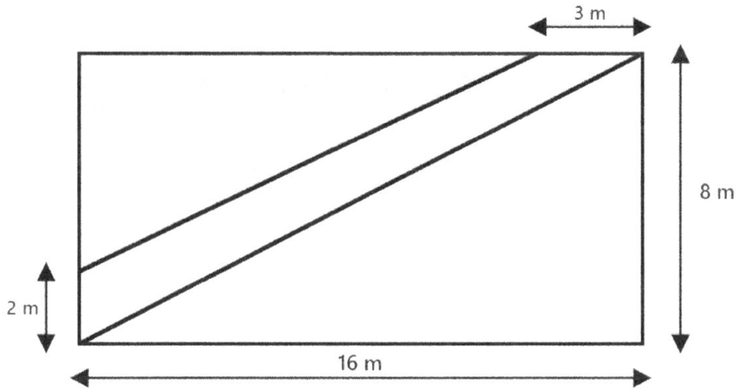

LET'S UNDERSTAND THE QUESTION TOGETHER

We have a **rectangle.**

A ribbon cuts across it and makes **two triangular sections**.

We need the area of the **smaller triangle**.

The side measurements in the image help us work out the triangle's real **base** and **height**.

The smaller triangle is the one in the **top-left corner**.

WHAT WE KNOW

- ✓ The rectangle is **16 m** wide.
- ✓ The rectangle is **8 m** high.
- ✓ The top marked part is **3 m.**
- ✓ The left marked part is **2 m**.
- ✓ The rectangle is divided into **two triangles.**
- ✓ We need the **smaller triangle.**

WHAT WE NEED TO FIND

The **area of the smaller triangular section**.

Our Plan

1. Find the triangle's **base** using the top side.
2. Find the triangle's **height** using the left side.
3. Use the triangle area formula.

Step-by-Step Solution

Step 1: Find the base of the smaller triangle.

The full width of the rectangle is **16 m**.

The top marked part on the right is **3 m**.

So, the top side of the smaller triangle is:

$$16 - 3 = 13$$

So, the base is **13 m**.

Step 2: Find the height of the smaller triangle.

The full height of the rectangle is **8 m**.

The marked part on the left from the bottom is **2 m**.

So, the height of the smaller triangle is:

$$8 - 2 = 6$$

So, the height is **6 m**.

Step 3: Use the area formula for a triangle.

$$\text{Area} = \frac{1}{2} \times \text{base} \times \text{height}$$

$$\text{Area} = \frac{1}{2} \times 13 \times 6$$

$$\text{Area} = 39$$

Quick Check

A triangle is always **half** of a matching rectangle with the same base and height.

If the base is **13 m** and the height is **6 m**, then the matching rectangle would have area:

$$13 \times 6 = 78$$

Half of **78** is **39**, so the answer is correct.

Final Answer

The area of the smaller triangular section is **39 m^2**.

Common Mistake

*A common mistake is to use **3 m** and **2 m** as the triangle's base and height.*

*But the smaller triangle uses the **remaining parts** of the sides, so we calculate:*

$16 - 3$ *and* $8 - 2$

Exam Tip

*Always ask yourself: **Which lengths belong to the triangle I want?** This helps you avoid using the wrong side measurements.*

Area Scale Factor (Squares)

QUESTION

A small square garden has an area of **45 square metres**.

A larger square garden has sides that are **4 times as long** as the sides of the small garden.

What is the area of the larger garden?

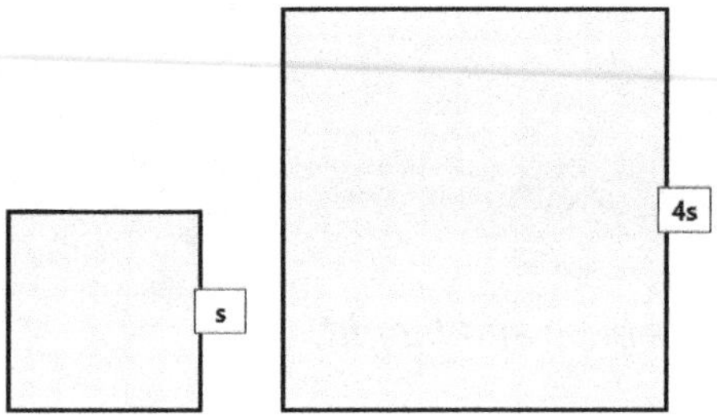

LET'S UNDERSTAND THE QUESTION TOGETHER

There are **two square gardens**.

he second garden is a **scaled-up version** of the first.

Its sides are **4 times as long** as the sides of the small garden.

The question is asking for the **area of the larger garden**.

WHAT WE KNOW

- ✓ The small square garden has an area of **45 m^2**
- ✓ The larger square has sides that are **4 times as long**

WHAT WE NEED TO FIND

The **area of the larger garden**.

Our Plan

When all lengths are multiplied by the same number, the area is multiplied by the square of that number.

So, we will:

- ✓ Find the area scale factor
- ✓ Multiply the original area by that factor

Step-by-Step Solution

Step 1: Find the length scale factor.

The sides are multiplied by **4**.

Step 2: Find the area scale factor.

$$4^2 = 16$$

So, the area is multiplied by **16**.

Step 3: Multiply the original area by 16.

$$45 \times 16 = 720$$

Step 4: Write the answer with units.

The area of the larger garden is **720 m²**.

Quick Check

If the sides are **4 times as long**, the area should be much bigger than **45 m²**.

Since the area is multiplied by **16**,

$$45 \times 16 = 720$$

This makes sense.

Final Answer

The area of the larger garden is **720 square metres**.

Common Mistake

A common mistake is to multiply ***45 by 4****.*

But when lengths are multiplied by ***4****, the area is multiplied by* $4^2 = 16$.

Exam Tip

This same rule applies to any similar shape, such as ***squares, rectangles, triangles, parallelograms and circles****.*

If all ***lengths*** *are multiplied by the same number, the area is multiplied by the* ***square*** *of that number.*

For circles, use the ***radius*** *or* ***diameter*** *instead of sides.*

Area Reduction in Triangles

QUESTION

A triangular cake has an area of **180 square centimetres**.

Nai makes a smaller **similar triangular slice of pudding**. Each side of the smaller triangle is **one third** of the length of the matching side of the cake.

What is the area of the smaller triangular slice of pudding?

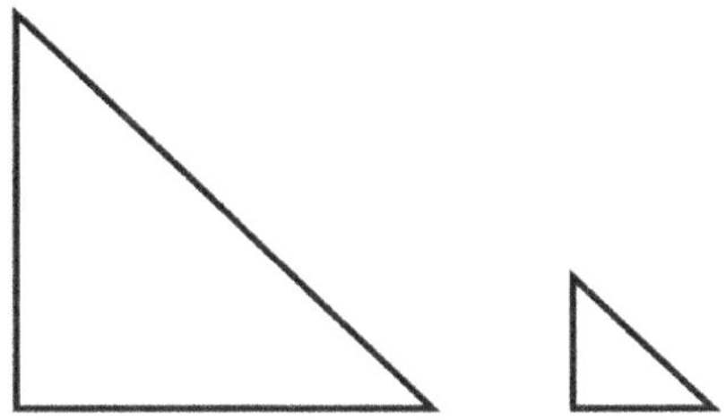

LET'S UNDERSTAND THE QUESTION TOGETHER

We start with a **triangle** that has an area of **180 square centimetres**.

Then a **smaller similar triangle** is made.

Similar means the new triangle has the same shape, but it is a different size.

The important detail is that **each side of the smaller triangle is one third of the original side length**.

The question is asking for the **area of the smaller triangle**, not the side lengths.

WHAT WE KNOW

- ✓ The original shape is a **triangle**.
- ✓ The original area is **180 cm^2**.
- ✓ The smaller triangle is **similar** to the original triangle.
- ✓ Each side of the smaller triangle is **one third** of the original side length.

WHAT WE NEED TO FIND

We need to find the **area of the smaller triangular slice of pudding**.

Our Plan

For **similar shapes**, when the side lengths change, the area changes by the **square of the scale factor**.

So, we will:

1. Find the side-length scale factor.
2. Square it to get the area scale factor.
3. Use that to find the new area.

Step-by-Step Solution

Step 1: Write the side-length scale factor.

Each side of the smaller triangle is **one third** $\left(\frac{1}{3}\right)$ of the original.

Step 2: Square the scale factor to find the area scale factor.

For similar shapes, area changes by:

$$\left(\frac{1}{3}\right)^2 = \frac{1}{9}$$

So, the smaller triangle has **one ninth** of the original area.

Step 3: Find one ninth of 180.

$$180 \div 9 = 20$$

Step 4: Write the answer with units. 20 cm^2

So, the area of the smaller triangular slice of pudding is **20 square centimetres**.

Quick Check

The side lengths became much smaller, so the area should also become much smaller.

Since the side-length scale factor is: $\frac{1}{3}$

the area scale factor should be:

$$\frac{1}{3} \times \frac{1}{3} = \frac{1}{9}$$

And: $180 \div 9 = 20$
So, the answer makes sense.

Final Answer

The area of the smaller triangular slice of pudding is **20 square centimetres**.

Common Mistake

*A common mistake is dividing **180 by 3** instead of by **9**.*

*When the **side lengths** change in similar shapes, the **area** changes by the square of that fraction.*

Exam Tip

*When you see **similar shapes**, remember:*

side scale factor first, then square it for area.

Area from Equal Perimeter

QUESTION

Nai baked a square cake and a rectangular cake. Both cakes have the same perimeter.

Use the diagram to work out the **area of the rectangular cake**.

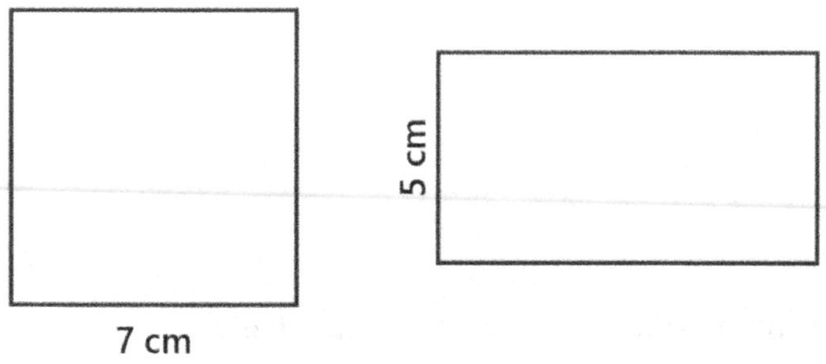

LET'S UNDERSTAND THE QUESTION TOGETHER

We are looking at **two cake shapes**.

They have the same **perimeter**, which means the total distance around each cake is equal.

The square cake has all sides the same, so we can work out its perimeter first.

The rectangular cake has one side given, but the **other side is missing**.

The question wants the **area of the rectangular cake**.

WHAT WE KNOW

- ✓ The square cake and the rectangular cake have the **same perimeter**.
- ✓ Each side of the square cake is **7 cm**.
- ✓ One side of the rectangular cake is **5 cm**.

WHAT WE NEED TO FIND

We need to find the **area of the rectangular cake**.

Our Plan

1. First, find the perimeter of the square cake.
2. Next, use that perimeter to find the missing side of the rectangular cake.
3. Then, multiply the two side lengths to find the area.

Step-by-Step Solution

Step 1: Find the perimeter of the square cake.
A square has 4 equal sides.

$\text{Perimeter} = 4 \times 7$

$= 28 \text{ cm}$

So, the rectangular cake also has a perimeter of **28 cm**.

Step 2: Set up the rectangle perimeter.

A rectangle has 2 pairs of equal sides.

So: $2 \times (5 + x) = 28$

Step 3: Divide both sides by 2.

$5 + x = 14$

Step 4: Find x.

$x = 14 - 5$
$x = 9$

So, the other side of the rectangular cake is **9 cm**.

Step 5: Find the area of the rectangular cake.

$\text{Area} = \text{length} \times \text{width}$
$= 5 \times 9$
$= 45 \text{ cm}^2$

Quick Check

Check the perimeter of the rectangular cake:

$$2 \times (5 + 9)$$
$$= 2 \times 14$$
$$= 28 \text{ cm}$$

This matches the square cake, so the answer is correct.

Final Answer

The area of the rectangular cake is **45 cm²**.

📌 Common Mistake

A common mistake is finding $\mathbf{x = 9}$ *and stopping.*

But the question asks for the **area**, *so you must keep going:*

Area = 5 × 9 = 45 cm²

Exam Tip

When two shapes have the ***same perimeter****, find the perimeter of the easier shape first.*

A ***square*** *works well because all 4 sides are equal.*
A ***rhombus*** *works the same way because it also has* ***4 equal sides****.*

Composite Area of Repeating Pattern

QUESTION

A wall pattern is made using a repeating block.

Each repeating block contains:

- ✓ One large square split into **4 equal small-square parts**
- ✓ One rectangle split into **2 equal small-square parts**

If the wall has a **4-by-7** pattern of these repeating blocks, what is the **total area of the wall pattern**?

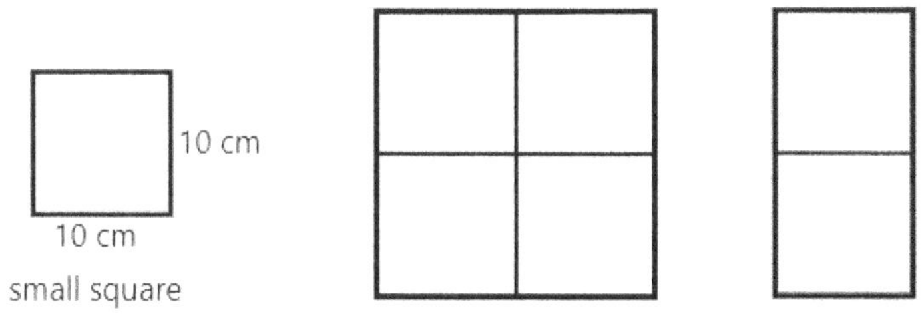

LET'S UNDERSTAND THE QUESTION TOGETHER

This question is about **area**, which means how much surface is covered.
We are finding the area of the shapes by comparing them to **small squares**.

- ✓ The large square has the same area as **4 small squares**
- ✓ The rectangle has the same area as **2 small squares**

So, **one repeating block** has the same area as:

4 small squares + 2 small squares = 6 small squares

A **4-by-7** pattern means **4 rows and 7 columns**, so there are **28 repeating blocks** altogether.

So, we need to:

1. Find the area of **one small square**.
2. Find the area of **one repeating block**.
3. Multiply by **28** to find the total area.

WHAT WE KNOW

- ✓ A small square is **10 cm × 10 cm**.
- ✓ Area of 1 small square = **100 cm²**.
- ✓ 1 large square has the same area as **4 small squares**.
- ✓ 1 rectangle has the same area as **2 small squares**.
- ✓ 1 repeating block has the same area as **6 small squares**.
- ✓ The pattern is **4 × 7**, so there are **28 blocks.**

WHAT WE NEED TO FIND

We need to find the **total area of the whole 4-by-7 pattern.**

Our Plan

1. First, find the area of **one small square**.
2. Then use that to find the area of **one repeating block**.
3. After that, find how many blocks there are in a **4-by-7 pattern**.
4. Finally, multiply to get the **total area**.

Step-by-Step Solution

Step 1: Find the area of one small square.

A small square is 10 cm × 10 cm.

$10 \times 10 = 100$

So, one small square has area **100 cm²**.

Step 2: Find the area of one repeating block.

Each repeating block is made from **6 small squares**.

$6 \times 100 = 600$

So, one repeating block has area **600 cm²**.

Step 3: Find the number of repeating blocks.

The pattern is **4 by 7**.

$4 \times 7 = 28$

So, there are **28 repeating blocks**.

Step 4: Find the total area.

Each block has area **600 cm²**.

$28 \times 600 = 16{,}800$

So, the total area of the wall pattern is **16,800 cm²**.

Quick Check

We can reverse-check by dividing the total area by the number of blocks:

$16{,}800 \div 28 = \mathbf{600}$

Since each block has area **600 cm²**, the answer is correct.

Final Answer

The total area of the 4-by-7 pattern is **16,800 cm²**.

Common Mistake

A common mistake is to find the area of ***one block*** *correctly but then forget to multiply by* ***28*** *for the whole pattern.*

Another common mistake is doing ***4 + 7*** *instead of* ***4 × 7*** *when finding the number of blocks.*

Exam Tip

When a question has a ***repeating pattern****, try to find the area of* ***one full block first****. Then count how many blocks there are, and multiply at the end. This keeps the working neat and helps you avoid mistakes.*

Composite Area (Square with Border Strips)

QUESTION

Mixi is designing a **square garden bed**. Around the edges, she places **4 identical rectangular garden strips**, leaving a **smaller square flower bed** in the middle.

Each rectangular garden strip is **2 m wide**.

The area of each rectangular garden strip is **16 m^2**.

What is the area of the small square flower bed in the centre?

LET'S UNDERSTAND THE QUESTION TOGETHER

The whole design makes **one large square garden bed.**

Around the edges are **4 identical rectangular strips.**

In the middle, there is a **smaller square flower bed.**

Because the strips are identical, they all have the same size.

We are told the **width** of each strip and the **area** of each strip.

The question wants the **area of the middle square flower bed.**

So, the important thing is to understand how the **rectangular strips fit around the square in the centre.**

WHAT WE KNOW

- ✓ Each rectangular garden strip is **2 m wide.**
- ✓ The area of each rectangular garden strip is **16 m^2**.
- ✓ All 4 rectangular strips are identical.
- ✓ The middle flower bed is a square.

WHAT WE NEED TO FIND

The **area of the square flower bed in the centre**.

Our Plan

First, find the missing length of each rectangular garden strip.

Next, notice that **one rectangle's length and another rectangle's width make the full outer side**.

Then, use that to find the side length of the large square.

After that, subtract the two strip widths to find the side length of the middle square.

Finally, multiply the side length of the middle square by itself to find its area.

Step-by-Step Solution

Step 1: Find the length of one rectangular garden strip.

Area of a rectangle = length × width

$16 \div 2 = 8$

So, the length of each rectangle is **8 m**.

Step 2: Write the dimensions of each rectangular strip.

Each strip is: 8 m × 2 m

Step 3: Find the side length of the large outer square.

In the design, one **rectangle's length** plus another **rectangle's width** makes the full outer side:

$8 + 2 = 10$

So, the large square has side length **10 m**.

Step 4: Find the side length of the centre square flower bed.

The middle square is what is left after the two strip widths are taken away:

$10 - 2 - 2 = 6$

So, the side length of the centre square is **6 m**.

Step 5: Find the area of the centre square flower bed.

Area of a square $=$ side $\times$ side

$$6 \times 6 = 36$$

So, the area of the centre square flower bed is: 36 m^2

Quick Check

The large square has side **10 m**.

Two strip widths take up **2 m + 2 m = 4 m**.

That leaves **6 m** for the middle square's side.

$$6 \times 6 = 36$$

So, the answer is correct.

Final Answer

The area of the small square flower bed is **36 m²**.

Common Mistake

*A common mistake is forgetting to remove **both** strip widths when finding the side length of the middle square.*

Exam Tip

*Before doing the final calculation, picture which lengths join together to make the **full side** of the larger shape. This can help you choose the correct numbers to add or subtract.*

Compassass Directions

QUESTION

Mika uses two identical right-angled isosceles triangles to make the shape shown.

She walks along the outside edges, starting at P, then going to Q, then to R, then to S.

What is the order of compass directions Mika follows?

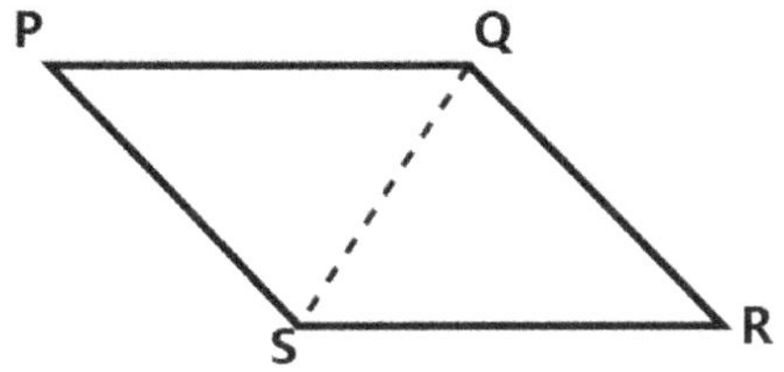

LET'S UNDERSTAND THE QUESTION TOGETHER

Mika is walking around the **outside edges** of the shape shown in the diagram.

She starts at **P**, then goes to **Q**, then **R**, then **S**.

We will use the picture to work out the direction of each part of her path:

- ✓ **P to Q**
- ✓ **Q to R**
- ✓ **R to S**

Since no special compass arrow is shown, we use the **usual compass directions**:

- ✓ **North** = Up
- ✓ **South** = Down
- ✓ **East** = Right
- ✓ **West** = Left

WHAT WE KNOW

- ✓ Mika walks **P → Q → R → S**.
- ✓ **PQ** is a horizontal side.
- ✓ **QR** is a slanted side going **down to the right**.
- ✓ **RS** is a horizontal side.
- ✓ We are using the **usual compass directions**.

WHAT WE NEED TO FIND

We need to find the order of compass directions Mika follows as she walks from **P** to **Q** to **R** to **S**.

Our Plan

We will:

- ✓ Find the direction from **P to Q**.
- ✓ Find the direction from **Q to R**.
- ✓ Find the direction from **R to S**.

Then we will write the directions in order.

Step-by-Step Solution

Step 1: Find the direction from P to Q.

From the diagram, **P to Q** is a horizontal line going to the **right**.

Right is **east**.

So, **P to Q = east**.

Step 2: Find the direction from Q to R.

From the diagram, **Q to R** goes **down to the right**.

Down and right is **south-east**.

So, **Q to R = south-east**.

Step 3: Find the direction from R to S.

From the diagram, **R to S** is a horizontal line going to the **left.**

Left is **west.**

So, **R to S = west.**

Step 4: Write the directions in order.

The directions are:

- ✓ **East**
- ✓ **South-east**
- ✓ **West**

Quick Check

Let's check the path against the diagram:

- ✓ **P to Q** goes right → **east.**
- ✓ **Q to R** goes down-right → **south-east.**
- ✓ **R to S** goes left → **west.**

These directions match the picture.

Final Answer

The order of compass directions Mika follows is **east, south-east, west.**

Common Mistake

*A common mistake is forgetting that **R to S** is the opposite direction of **S to R**.*
*From **R to S**, Mika moves **left**, so the direction is **west**.*

Exam Tip

When a question asks for directions along a path, label each small journey one at a time.
This makes it much easier to keep the directions in the correct order.

Lines of Symmetry (Quadrilateral)

QUESTION

Nai is thinking of a quadrilateral.

It has these features:

- ✓ **No sides are parallel.**
- ✓ **Two adjacent sides are 6 cm long.**
- ✓ **The other two sides are 9 cm long.**
- ✓ **Exactly one angle is a right angle.**

How many lines of symmetry does Nai's quadrilateral have?

LET'S UNDERSTAND THE QUESTION TOGETHER

A **quadrilateral** is a closed shape with **4 sides** and **4 angles.**

In this question, the quadrilateral has:

- ✓ **2 adjacent sides** that are **6 cm**
- ✓ **2 sides** that are **9 cm**
- ✓ **No parallel sides**
- ✓ **Exactly 1 right angle**

We are not being asked to draw the shape.

We are being asked to work out how many **lines of symmetry** it has.

A line of symmetry is a line that splits a shape into **2 matching mirror halves.**

So, we need to decide whether this quadrilateral could have **0, 1, 2, or more** lines of symmetry.

WHAT WE KNOW

1. It is a quadrilateral, so it has **4 sides**.
2. No sides are parallel.
3. Two sides are **6 cm** each.
4. Two sides are **9 cm** each.
5. It has **exactly one right angle**.

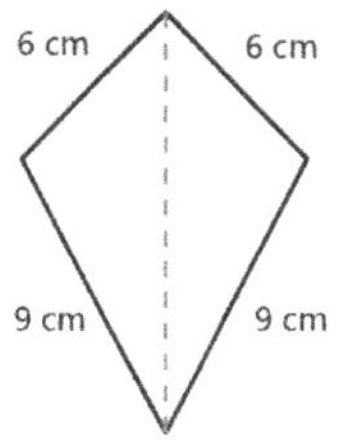

WHAT WE NEED TO FIND

How many **lines of symmetry** the quadrilateral has.

Our Plan

1. Think about the different numbers of symmetry lines a quadrilateral can have.
2. Check which of those are possible using the side lengths.
3. Rule out any that do not fit the parallel-side clue.
4. Check that the right-angle clue still works.
5. Choose the correct number of lines of symmetry.

Step-by-Step Solution

Step 1: Think about what a line of symmetry means.

A line of symmetry divides a shape into **2 matching mirror-image halves**.

Step 2: Recall the possible numbers of symmetry lines.

Different quadrilaterals can have different numbers of symmetry lines:

- ✓ **0 lines**: a general quadrilateral or an irregular trapezium.
- ✓ **1 line**: a kite or an isosceles trapezium.
- ✓ **2 lines**: a rectangle or a rhombus.
- ✓ **4 lines**: a square.

So, we need to work out which number fits this quadrilateral.

Step 3: Use the side lengths.

The side lengths come in **two equal pairs**:

- ✓ 6 cm and 6 cm
- ✓ 9 cm and 9 cm

Since the equal sides are adjacent, the quadrilateral must be a kite. A kite has 1 line of symmetry.

If we draw a quadrilateral with these side lengths and one line of symmetry, it will look like a **kite**.

Step 4: Use the parallel-side clue.

Quadrilaterals with **2** or **4** lines of symmetry have **parallel sides**.

But this question says **no sides are parallel**.

So, this quadrilateral cannot have **2** or **4** lines of symmetry.

Step 5: Check the right-angle clue.

The shape has **exactly one right angle**.

This still fits a quadrilateral with **1 line of symmetry**.

So, the right-angle clue does not change our answer.

Step 6: State the answer.

The quadrilateral has **1 line of symmetry**.

Quick Check

- ✓ The equal side lengths make **1 line of symmetry possible**.
- ✓ **No parallel sides** rules out **2** or **4** lines of symmetry.
- ✓ The **one right angle** still fits.

So, **1 line of symmetry** is correct.

Final Answer

Nai's quadrilateral has **1 line of symmetry**.

Common Mistake

*A common mistake is to think a quadrilateral must always have the same number of symmetry lines. But different quadrilaterals can have **0, 1, 2, or 4** lines of symmetry.*

Exam Tip

Use each clue one at a time. First check the side lengths, then the parallel sides, then the angle clue. This helps you rule out wrong answers carefully.

Reflections of a Square Tile

QUESTION

A square tile has corners labelled A, B, C, and D in order: A top left, B top right, C bottom right, and D bottom left.

The tile is flipped in this order:

1. Across its **vertical line of symmetry**.
2. Across its **horizontal line of symmetry**.
3. Across its **vertical line of symmetry** again.

After all the flips, **which label is in the top left corner of the square?**

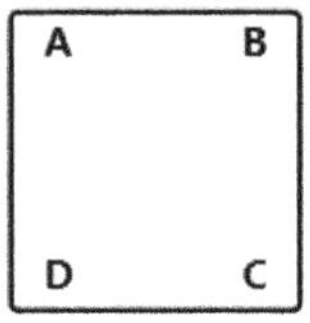

LET'S UNDERSTAND THE QUESTION TOGETHER

We are starting with a **square** that has one label at each corner.

The corners begin in a clear order around the square:

- ✓ Top left is **A**
- ✓ Top right is **B**
- ✓ Bottom right is **C**
- ✓ Bottom left is **D**

Then the square is flipped **three times**.

Each flip changes where the labels go.

We are not being asked to draw anything fancy. We only need to work out **which label finishes in the top left corner** at the end.

WHAT WE KNOW

- ✓ The starting corners are:
  - Top left = **A**
  - Top right = **B**
  - Bottom right = **C**
  - Bottom left = **D**
- ✓ First flip: across the **vertical** line of symmetry
- ✓ Second flip: across the **horizontal** line of symmetry
- ✓ Third flip: across the **vertical** line of symmetry again

WHAT WE NEED TO FIND

We need to find **which label ends up in the top left corner** after all 3 flips.

Our Plan

We will track the corner labels **one flip at a time**.

After each flip, we will rewrite where the labels move.

Then we will look at the final **top left** corner.

Step-by-Step Solution

Step 1: Write the starting positions.

- ✓ Top left = **A**
- ✓ Top right = **B**
- ✓ Bottom right = **C**
- ✓ Bottom left = **D**

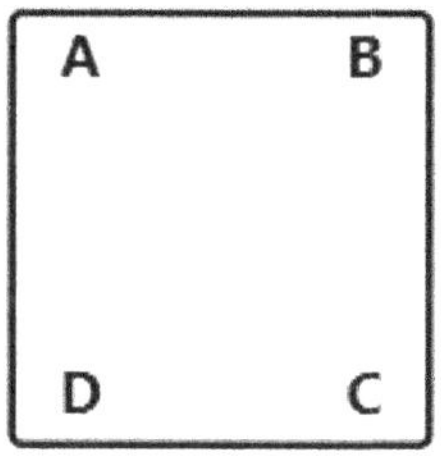

Step 2: Do the first flip across the vertical line of symmetry.

A reflection across the vertical line swaps left and right.

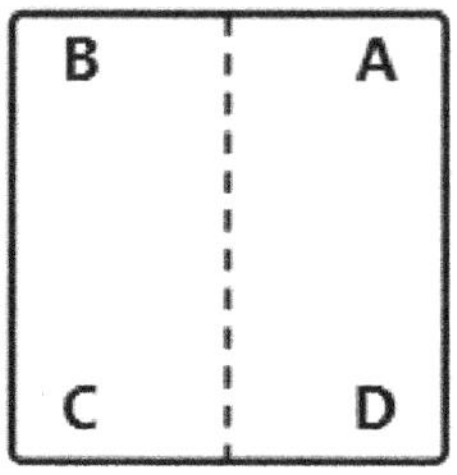

So:

- ✓ Top left and top right swap.
- ✓ Bottom left and bottom right swap.

New positions:

- ✓ Top left = **B**
- ✓ Top right = **A**
- ✓ Bottom right = **D**
- ✓ Bottom left = **C**

Step 3: Do the second flip across the horizontal line of symmetry.

A reflection across the horizontal line swaps top and bottom.

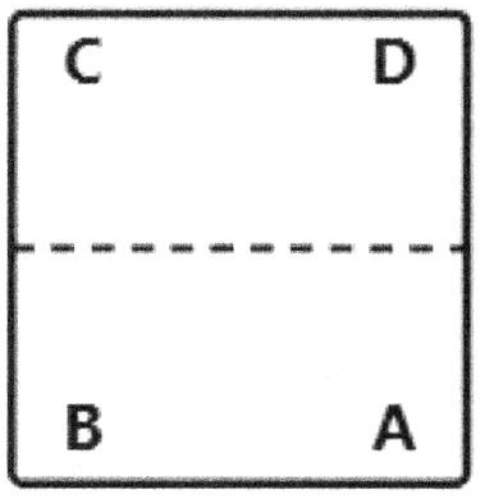

So:

- ✓ Top left and bottom left swap.
- ✓ Top right and bottom right swap.

New positions:

- ✓ Top left = **C**
- ✓ Top right = **D**
- ✓ Bottom right = **A**
- ✓ Bottom left = **B**

Step 4: Do the third flip across the vertical line of symmetry again.

Again, a vertical flip swaps **left and right**.

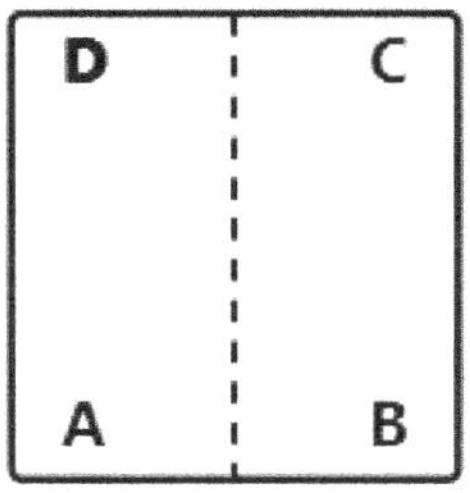

So:

- ✓ Top left and top right swap.
- ✓ Bottom left and bottom right swap.

New positions:

Top left = **D**, Top right = **C**, Bottom right = **B**, Bottom left = **A**.

Quick Check

Let's check the movement of **D**:

- ✓ Starts at bottom left.
- ✓ After first vertical flip, it goes to bottom right.
- ✓ After horizontal flip, it goes to top right.
- ✓ After final vertical flip, it goes to top left.

That matches our answer.

Final Answer

The label in the top left corner after the reflections is **D**.

Common Mistake

A common mistake is to do all the flips in your head at once. That can mix up the corners. It is much safer to track the labels ***after each single flip****.*

Exam Tip

When a shape is reflected more than once, make a tiny corner list after every move. Keeping the order neat helps you avoid losing track of the labels.

Metric Distance Conversion Errors

QUESTION

A hiking trail sign shows that two lookout points are **6 kilometres 45 metres** apart.

Nai accidentally writes the distance as **45 kilometres 6 metres**.

What is the difference between the correct distance and the incorrect distance?

LET'S UNDERSTAND THE QUESTION TOGETHER

We are comparing **two distances**.

One distance is the **correct** one: **6 km 45 m**.

The other distance is the **incorrect** one: **45 km 6 m**.

They use the same numbers, but the **kilometres and metres have been mixed up**.

We need to work out **how much bigger one distance is than the other**.

WHAT WE KNOW

- ✓ Correct distance = **6 km 45 m**
- ✓ Incorrect distance = **45 km 6 m**
- ✓ We need to find the **difference** between them.

WHAT WE NEED TO FIND

We need to find the difference between the two distances.

Our Plan

Change both distances into **metres** first.

Then subtract the smaller distance from the larger distance.

Step-by-Step Solution

Step 1: Change the correct distance into metres.

$6 \text{ km} = 6000 \text{ m}$
So,
$6 \text{ km } 45 \text{ m} = 6000 + 45 = 6045 \text{ m}$

Step 2: Change the incorrect distance into metres.

$45 \text{ km} = 45000 \text{ m}$
So,
$45 \text{ km } 6 \text{ m} = 45000 + 6 = 45006 \text{ m}$

Step 3: Subtract to find the difference.

$45006 - 6045 = 38961$

So, the difference is **38,961 m**.

Step 4: Change the answer back into kilometres and metres.

$38961 \text{ m} = 38 \text{ km } 961 \text{ m}$

Quick Check

The incorrect distance is about **45 km**, and the correct distance is about **6 km**.

The difference should be about **39 km**.

Our answer, **38 km 961 m**, is very close to **39 km**, so it makes sense.

Final Answer

The difference between the two distances is **38 km 961 m**.

Common Mistake

*A common mistake is to subtract the kilometres and metres separately without first checking the units carefully. It is much safer to change both distances into **metres** first.*

Exam Tip

*When a question uses **kilometres and metres**, always make the units the same before you calculate. This helps prevent simple mistakes.*

Overlapping Lengths in a Line

QUESTION

Three identical garden paths are laid in a straight line.
The second path covers part of the end of the first path, and the third path covers part of the end of the second path. The total distance from the start of the first path to the end of the third path is shown in the image.

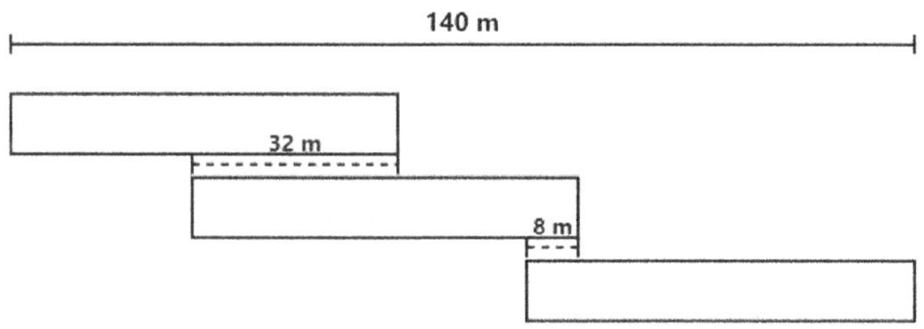

What is the length of one garden path?

LET'S UNDERSTAND THE QUESTION TOGETHER

There are **3 equal paths.**

They are placed in a straight line, but they **overlap**, so the whole distance is not just 3 full lengths added together.

We will use the measurements shown in the diagram.

We need to find the length of **one** path.

WHAT WE KNOW

There are **3 identical** garden paths.

The overlap measurements are shown in the image.

The total distance is shown in the image.

All paths have the **same length.**

WHAT WE NEED TO FIND

The length of **one** garden path.

Our Plan

1. We will use the measurements shown in the image.
2. Then we will add the 3 equal path lengths and subtract the overlaps.
3. After that, we will solve the equation.

Step-by-Step Solution

Let each garden path have length L **m**.

Step 1: Start with the first path

The first path gives the full length: $\boldsymbol{L}$

Step 2: Add the extra part of the second path

The second path overlaps the first by **32 m**, so it does not add its full length.

It adds: $\boldsymbol{L - 32}$

Step 3: Add the extra part of the third path

The third path overlaps the second by **8 m**, so it adds: $\boldsymbol{L - 8}$

Step 4: Write, simplify, and solve the equation

*The total distance is **140 m**, so:*

$$L + (L - 32) + (L - 8) = 140$$
$$3L - 40 = 140$$
$$3L = 180$$
$$L = 60$$

Quick Check

Subtract the overlaps: $\mathbf{180 - 32 - 8 = 140}$

Final Answer

One garden path is **60 m** long.

📌 Common Mistake

A common mistake is forgetting to subtract the overlapping parts.

💡 Exam Tip

*When shapes or objects **overlap**, do not add every full length without thinking.*

Rectangular Shadows of 3D Shapes

QUESTION

Nai is testing some solid shapes with a torch in a dark room.

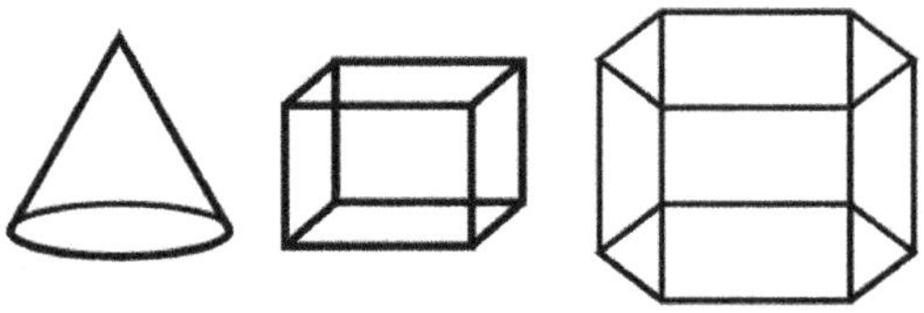

Nai turns each shape in different ways to make shadows on a wall.

Which of these shapes **can make a rectangular shadow**? Write the name of each shape that works.

LET'S UNDERSTAND THE QUESTION TOGETHER

We are looking at **3D shapes** and the **shadows** they can make.

The torch can shine from **different directions**, so the shape may be turned around.

That means we are not just thinking about the shape from one fixed view. We are thinking about **all possible views.**

We need to decide which shapes could show a **rectangle-shaped shadow** on the wall.

WHAT WE KNOW

Nai is using:

- A cone.
- A rectangular prism.
- A hexagonal prism.

The torch can shine from **different directions.**

We need to find which shapes can make a **rectangular shadow.**

WHAT WE NEED TO FIND

We need to find **all the shapes** that can produce a rectangular shadow.

Our Plan

We will think about what each 3D shape can look like when light shines on it from a suitable direction.

Step-by-Step Solution

Step 1: Think about the **cone**.

A cone has a curved surface and a circular base.
Its shadows can look like a triangle or a circle, depending on the direction.

So, the cone does **not** make a rectangular shadow.

Step 2: Think about the **rectangular prism**.

A rectangular prism has rectangular faces.
If the light shines straight onto one of those faces, the shadow can be a **rectangle**.

So, the rectangular prism **can** make a rectangular shadow.

Step 3: Think about the **hexagonal prism**.

A hexagonal prism has hexagon ends, but its side faces are rectangles.

If the hexagonal prism is viewed from the side, its outline can be a rectangle.

So, the hexagonal prism **can** make a rectangular shadow.

Step 4: Write the shapes that work.

The shapes that can make a rectangular shadow are:

✓ Rectangular prism.

✓ Hexagonal prism.

Quick Check

A shape can make a rectangular shadow if it has a **rectangular face** that can face the light, or if its side view can look like a rectangle.

- ✓ Cone: no.
- ✓ Rectangular prism: yes.
- ✓ Hexagonal prism: yes.

So, our answer makes sense.

Final Answer

The shapes that can make a rectangular shadow are the **rectangular prism** and the **hexagonal prism**.

📌 Common Mistake

*A common mistake is thinking only the **name of the end face** matters.*
*For example, a hexagonal prism has hexagon ends, but its **side faces are rectangles**, so it can still make a rectangular shadow.*

Exam Tip

When light shines from different directions, imagine turning the 3D shape.

A shape can make a rectangular shadow if:

- ✓ *It has a rectangular face, or*
- ✓ *It can appear rectangular from a side view.*

*Always think about the **outline (silhouette)** of the shape, not just the shape of its ends.*

Properties of 3D Shapes

QUESTION

Nai is making a **3D shape fact card** for these 5 shapes:

- ✓ Cube.
- ✓ Rectangular prism.
- ✓ Triangular prism.
- ✓ Square pyramid.
- ✓ Cylinder.

Using the table, Nai writes down the number of **faces**, **edges**, and **vertices** for each shape.

Then Nai adds each group together.

What is the **total number of faces**, **total number of edges**, and **total number of vertices** for all 5 shapes combined?

LET'S UNDERSTAND THE QUESTION TOGETHER

We are looking at **5 different 3D shapes.**

Each shape has:

- ✓ A number of **faces**.
- ✓ A number of **edges**.
- ✓ A number of **vertices**.

The question wants us to find **3 totals**:

- ✓ All the faces added together.
- ✓ All the edges added together.
- ✓ All the vertices added together.

We are adding the values for all 5 shapes.

WHAT WE KNOW

- ✓ **Cube:** 6 faces, 12 edges, 8 vertices.
- ✓ **Rectangular prism:** 6 faces, 12 edges, 8 vertices.
- ✓ **Triangular prism:** 5 faces, 9 edges, 6 vertices.
- ✓ **Square pyramid:** 5 faces, 8 edges, 5 vertices.
- ✓ **Cylinder:** 3 faces (including 1 curved face), 2 edges, 0 vertices.

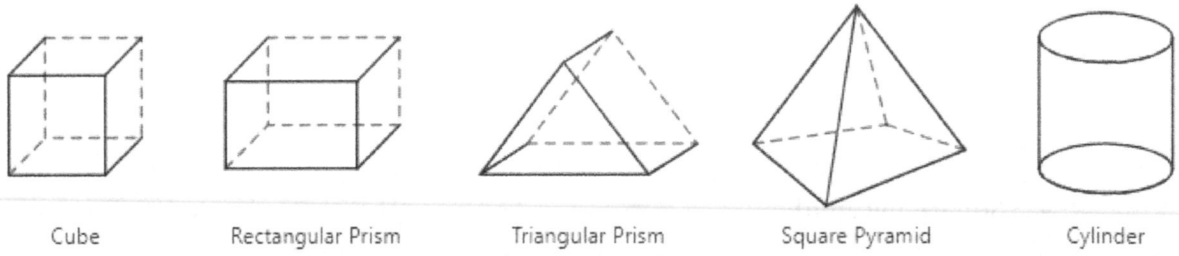

WHAT WE NEED TO FIND

We need to find:

- ✓ The **total number of faces**.
- ✓ The **total number of edges**.
- ✓ The **total number of vertices**.

for all 5 shapes together.

Our Plan

We will:

1. Add all the **faces**.
2. Add all the **edges**.
3. Add all the **vertices**.
4. Write the three totals clearly.

Step-by-Step Solution

Step 1: Add the faces

Faces: Cube + Rectangular prism + Triangular prism + Square pyramid + Cylinder
$= 6 + 6 + 5 + 5 + 3$
$= 25$

So, the total number of faces is **25**.

Step 2: Add the edges

Edges: Cube + Rectangular prism + Triangular prism + Square pyramid + Cylinder
$= 12 + 12 + 9 + 8 + 2$
$= 43$

So, the total number of edges is **43**.

Step 3: Add the vertices

Vertices: Cube + Rectangular prism + Triangular prism + Square pyramid + Cylinder
$= 8 + 8 + 6 + 5 + 0$
$= 27$

So, the total number of vertices is **27**.

Quick Check

Let's check by thinking carefully:

- ✓ Most shapes have 5 or 6 faces, so for 5 shapes a total near **25** makes sense.
- ✓ Edges are mostly between 8 and 12, so a total of **43** sounds reasonable.
- ✓ Vertices are mostly between 0 and 8, so **27** is also sensible.

Our totals fit the table well.

Final Answer

The 5 shapes have **25 faces, 43 edges, and 27 vertices** altogether.

Common Mistake

A common mistake is to forget one of the shapes when adding, especially the ***cylinder****, or to mix up* ***edges*** *and* ***vertices****.*

Exam Tip

When a question asks for several totals, work in neat groups. Add ***all faces first****, then* ***all edges****, then* ***all vertices*** *so nothing gets mixed up.*

Painted Cube Face Counting

QUESTION

A large solid cube is made using **125 identical small cubes**.

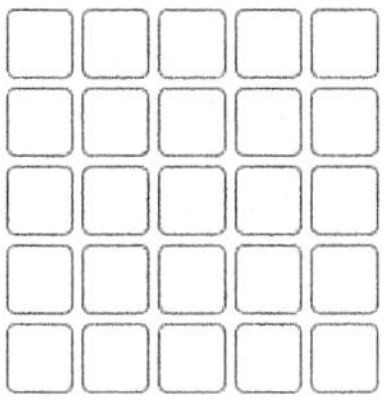

The cube is painted **grey on all outer faces**, then **completely dried**.

After that, a smaller cube of size **$3 \times 3 \times 3$** is removed from the **exact centre** of the large cube.

How many of the remaining small cubes have **exactly two faces painted grey**?

LET'S UNDERSTAND THE QUESTION TOGETHER

The large cube is made from **125 small cubes**.

Since $125 = 5 \times 5 \times 5$, the large cube is a $5 \times 5 \times 5$ **cube**.

It is painted on the **outside only**, so only cubes on the outer surface get **grey** paint.

Then a $3 \times 3 \times 3$ cube is removed from the **exact centre**.

That means the outside painted surface stays the same, but the inner centre cubes are removed.

We need to find how many **remaining** small cubes have **exactly 2 faces painted grey**.

WHAT WE KNOW

- ✓ The large cube is $5 \times 5 \times 5$
- ✓ The outer faces are painted grey.
- ✓ A $3 \times 3 \times 3$ cube is removed from the centre.
- ✓ We want cubes with **exactly 2 grey faces.**

WHAT WE NEED TO FIND

The number of small cubes that still remain and have **exactly two outer painted faces.**

Our Plan

Cubes with exactly 2 painted faces are the cubes along the **edges**, but **not the corners**.

So, we will count those edge cubes and then check whether removing the centre changes them.

Step-by-Step Solution

Step 1: Work out the size of the large cube

The large cube is made from **125** identical small cubes.

Since $\mathbf{125 = 5 \times 5 \times 5}$, the large cube has side length **5** small cubes.

Large cube $= 5 \times 5 \times 5$

Step 2: Understand which cubes can have exactly 2 painted faces

Only small cubes on the **edges** of the big cube can have exactly **2** painted faces.

Corner cubes have **3** painted faces, and face-middle cubes have only **1** painted face.

So, we only need to count the small cubes that sit on edges but **not at the corners**.

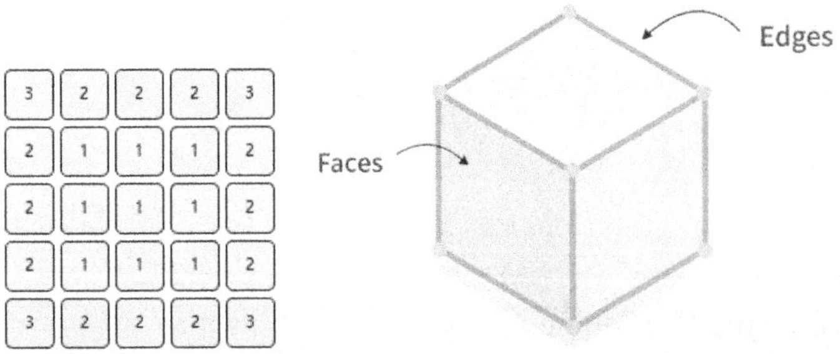

Step 3: Count the edge cubes with exactly 2 painted faces

Along each edge of a **5 × 5 × 5** cube, the two end cubes are corners, so they do **not** count.

That leaves **5 - 2 = 3** cubes on each edge with exactly 2 painted faces.

A cube has **12 edges**. So total cubes with exactly 2 painted faces: $12 \times 3 = 36$

Step 4: Check whether removing the $3 \times 3 \times 3$ centre changes this

The cube was painted **before** the centre was removed.

That means only the **outside surface** got painted grey.

When the centre **3 × 3 × 3** part is taken out, the newly exposed inside faces are **not painted.**

So, the cubes that already had exactly 2 painted faces stay the same in number.

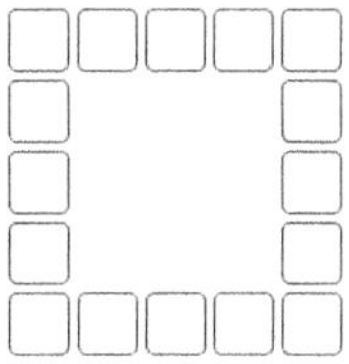

Quick Check

Edge cubes with 2 painted faces are always on the outside edges only. Since the centre is removed, those outside edge cubes stay untouched. So, **36** makes sense.

Final Answer

36 small cubes have exactly two faces painted grey.

Common Mistake

A common mistake is to think the inner hole creates new ***grey*** *faces. It does not. Only the* ***original outer faces*** *were painted grey.*

Exam Tip

In any painted cube: ***corner = 3 faces, edge = 2 faces, face-centre = 1 face, inside = 0 faces.***

Edge Painted Cube Counting

QUESTION

Mika builds a large cube using identical small cubes.

There are **5 small cubes along each edge** of the large cube.

A strip of black paint is brushed **only along every outside edge** of the large cube.

How many small cubes have black paint on them?

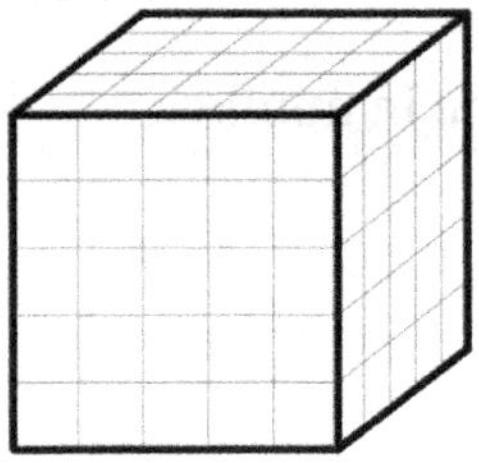

LET'S UNDERSTAND THE QUESTION TOGETHER

We are looking at a **large cube** made from many **small cubes**.

The paint is brushed **only along the outside edges** of the large cube.

So, a small cube will have paint on it only if it sits on one of those **outside edges**.

We need to find the total number of small cubes that touch **at least one edge** of the large cube.

WHAT WE KNOW

- ✓ The large cube is made from identical small cubes.
- ✓ There are **5 small cubes along each edge**.
- ✓ A cube has **12 edges**.
- ✓ We want the number of small cubes that have paint on them.

WHAT WE NEED TO FIND

The number of small cubes that lie on the **outside edges** of the large cube.

Our Plan

Use the formula $\mathbf{12n - 16}$, where **n** is the number of small cubes along one edge.

Step-by-Step Solution

Step 1: Identify n.

There are 5 small cubes on each edge.
So, $\mathbf{n = 5}$.

Step 2: Use the formula for cubes on the edges.

Cubes on the edges = $\mathbf{12n - 16}$

Step 3: Substitute $n = 5$ into the formula.

$12(5) - 16$
$= 60 - 16$
$= 44$

So, **44** small cubes have black paint on them.

Quick Check

Take away the 8 corner cubes from 44: $\mathbf{44 - 8 = 36}$

Now divide by the 12 edges: $\mathbf{36 \div 12 = 3}$

So, each edge has **3 middle cubes**.

Add the 2 corner cubes at the ends: $\mathbf{3 + 2 = 5}$

That matches the question, so **44 is correct**.

Final Answer

44 small cubes have black paint on them.

Common Mistake

A common mistake is to do: $12 \times 5 = 60$ *and stop there.*

*This counts the **corner cubes too many times**, because each corner sits on **3 edges**.*

Exam Tip

*For a cube with **n** small cubes along each edge:*

- ✓ *Total small cubes* $= n^3$
- ✓ *Corner cubes* $= 8$
- ✓ *Cubes on the surface* $= 6n^2 - 12n + 8$
- ✓ *Hidden inside cubes* $= (n-2)^3$
- ✓ *Cubes on the edges* $= 12n - 16$

Interior Cubes in a Cube

QUESTION

Mika builds a big cube using **125 identical small cubes**.

She turns the cube around and looks at it from **all sides**.

How many of the small cubes are completely hidden inside the big cube and **cannot be seen at all**?

LET'S UNDERSTAND THE QUESTION TOGETHER

The big cube is made by joining lots of **small cubes**.

Because Mika looks from **all sides**, every cube on the **outside surface** can be seen.

So, we are not counting cubes on the top, bottom, front, back, left, or right faces.

We are only counting the cubes that are **fully inside** the big cube.

These inside cubes are hidden because they are covered by outer cubes on every side.

WHAT WE KNOW

- ✓ The total number of small cubes is **125**.
- ✓ The solid shape is a **cube**.
- ✓ Mika can look at it from **any direction**.
- ✓ Any cube on the **outside** can be seen.
- ✓ Only the cubes in the **inner part** cannot be seen.

WHAT WE NEED TO FIND

We need to find the number of small cubes that are **completely inside** the large cube.

Our Plan

We will first find how many small cubes are along one edge of the large cube.

Then we will use the inside-cube formula:

$$(\boldsymbol{n} - \mathbf{2})^3$$

where $\boldsymbol{n}$ is the number of small cubes along one edge.

Step-by-Step Solution

Step 1: Find the number of small cubes along one edge.

Since the total number of cubes is **125**, and

$$125 = 5^3$$

The large cube has 5 small cubes along each edge.

So $n = 5$.

Step 2: Write the formula for **hidden cubes.**

For a cube with $\boldsymbol{n}$ small cubes along each edge, the number of cubes completely hidden inside is:

$$(\boldsymbol{n} - \mathbf{2})^3$$

This works because we remove **one outside layer** from each side of the cube.

Step 3: Substitute $n = 5$into the formula.

$= (5-2)^3$

$= (3)^3$

$= 27$

So, **27 small cubes** are completely hidden inside the large cube and cannot be seen at all.

Quick Check

Let's check using logic.

A $\mathbf{5 \times 5 \times 5}$ cube has one full outside layer.

After removing that outside layer from every side, the inside part is:

$$\mathbf{3 \times 3 \times 3 = 27}$$

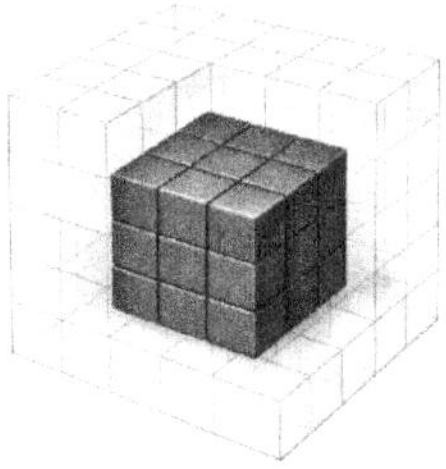

Final Answer

There are **27 small cubes** that cannot be seen at all.

Common Mistake

A common mistake is thinking that all the cubes you cannot see in ***one picture*** *are hidden.*

But Mika turns the cube and looks from ***every direction****, so any cube on the outside will be seen at some point.*

Only the cubes right in the ***middle*** *are completely hidden.*

Exam Tip

In cube questions, first find the ***edge length*** *of the large cube.*

Then imagine peeling off the ***outside layer*** *all around.*

The cubes left in the centre are the hidden ones.

Hollow Cube Counting (face views)

QUESTION

A large cube is made from **64 identical small cubes**.

Some of the small cubes are removed. After that, whenever Nai looks **straight at any one of the 6 faces**, the face appears as a **4 by 4 square border with a 2 by 2 empty hole in the centre**.

How many small cubes are left in the shape?

LET'S UNDERSTAND THE QUESTION TOGETHER

The shape started as one **solid cube** made from small cubes.

Since there are **64 small cubes**, the large cube must be **4 cubes long, 4 cubes wide, and 4 cubes high**.

After some cubes are removed, every face shows the same pattern:

- ✓ Cubes around the outside.
- ✓ An **empty middle**.
- ✓ The empty middle is **2 by 2**

Because this happens on **all 6 faces**, the missing part is not just on one face. The shape has been hollowed through the middle so that only the outside edge frame remains.

We need to find **how many cubes are still left**.

WHAT WE KNOW

- ✓ The large cube has **64 small cubes**.
- ✓ $64 = 4^3$, so, the cube is $\mathbf{4 \times 4 \times 4}$.
- ✓ Each face looks like a **square border**.
- ✓ The middle $\mathbf{2 \times 2}$ **section** of each face is empty.
- ✓ This happens on **all 6 faces**.

WHAT WE NEED TO FIND

We need to find the **number of small cubes left** in the object.

Our Plan

We will recognise that the remaining cubes form the **edge frame** of the cube. Then we will use the formula $\mathbf{12n - 16}$ to count the cubes quickly.

Step-by-Step Solution

Step 1: Find the side length of the large cube

The cube is made from 64 small cubes.

$$\mathbf{64 = 4^3}$$

So, the side length is: $\mathbf{n = 4}$

Step 2: Identify the remaining shape

Since every face shows a hollow square border, the cubes left are the cubes on the **edges** of the large cube.

So, the remaining shape is the **cube frame**.

Step 3: Use the cube-frame formula

For a cube with side length n, the number of cubes in the frame is:

$12n - 16$

Step 4: Substitute $n = 4$

$$12(4) - 16$$

$$12 \times 4 = 48$$

$$48 - 16 = 32$$

So, the number of cubes left is: 32

Quick Check

Let us check another way.

A cube has:

- ✓ **12 edges**
- ✓ **8 corners**

If each edge has 4 cubes, then: $12 \times 4 = 48$

But this counts each corner **3 times**, because each corner belongs to 3 edges.

So, we must subtract the extra corner counts.

Each of the 8 corners has been over-counted by 2, so subtract:

$$8 \times 2 = 16$$

Now we get:

$$48 - 16 = 32$$

So, the answer is correct.

Final Answer

There are **32 small cubes** left in the shape.

Common Mistake

A common mistake is to count all the cubes along the 12 edges as $12n$ *and forget that the* ***corner cubes are counted too many times****. Each corner belongs to* ***3 edges****, so the corners need to be adjusted when finding the total.*

Exam Tip

For a cube frame with side length n*, a very useful shortcut is:*

$12n - 16$

It works because you count all ***12 edges****, then fix the* ***over-counted corners****.*

Cube with Intersecting Tunnels

QUESTION

Mika builds a solid cube using 1 cm cubes. The large cube has side length 6 cm.

She then cuts out three square tunnels, each with side length 4 cm, so that:

- ✓ One tunnel goes straight from front to back.
- ✓ One tunnel goes straight from left to right.
- ✓ One tunnel goes straight from top to bottom.

Each tunnel goes all the way through the cube and passes through the middle.

How many 1 cm cubes are left after all three tunnels are removed?

LET'S UNDERSTAND THE QUESTION TOGETHER

We begin with a **full cube** that is **6 cm long on every edge**.

Because it is made from **1 cm cubes**, the whole shape is built from lots of little cubes.

Then **three square tunnels** are cut through the middle.

Each tunnel:

- ✓ Is **4 cm by 4 cm**
- ✓ Goes all the way through the cube.
- ✓ Passes through the **centre**.

So, the middle of the cube is hollowed out in **three directions**.

We need to work out how many small cubes are **still left** after all that cutting.

WHAT WE KNOW

- ✓ The large cube has side length **6 cm.**
- ✓ It is made from **1 cm cubes**.
- ✓ There are **3 tunnels**.
- ✓ Each tunnel has square side length **4 cm.**
- ✓ Each tunnel passes through the **middle** of the cube.

WHAT WE NEED TO FIND

We need to find how many **1 cm cubes are left** after the three tunnels are removed.

Our Plan

We will use the pattern $\boldsymbol{12n - 16}$ for the number of cubes left.

Then we will substitute $\boldsymbol{n = 6}$ because the cube has side length 6 cm.

After that, we will calculate the final number of cubes left.

Step-by-Step Solution

Step 1: Write the formula

$12n - 16$

Step 2: Substitute $n = 6$and calculate

$12(6) - 16$

$= 72 - 16$

$= 56$

So, the number of cubes left is: 56

Quick Check

$12 \times 6 = 72$, and $72 - 16 = 56$.

The answer is a reasonable size because a lot of the middle has been removed, leaving just the outer frame-like parts.

Final Answer

There are **56** one-centimetre cubes left.

Common Mistake

Some students try to calculate the total number of cubes first (like $6 \times 6 \times 6$*) and then subtract tunnels. This becomes messy and easy to get wrong.*

Instead, use the pattern $12n - 16$ *directly to save time and avoid errors.*

Exam Tip

If you recognise a known pattern or formula, ***use it straight away****.*

Write the formula first, then substitute carefully in one clean line:

$$12(6) - 16 = 72 - 16 = 56$$

Neat working helps you think clearly and reduces mistakes.

Unit Cubes in a Rectangular Block

QUESTION

Nai makes a solid rectangular block using identical unit cubes.

From three different sides, the block shows these face dimensions:

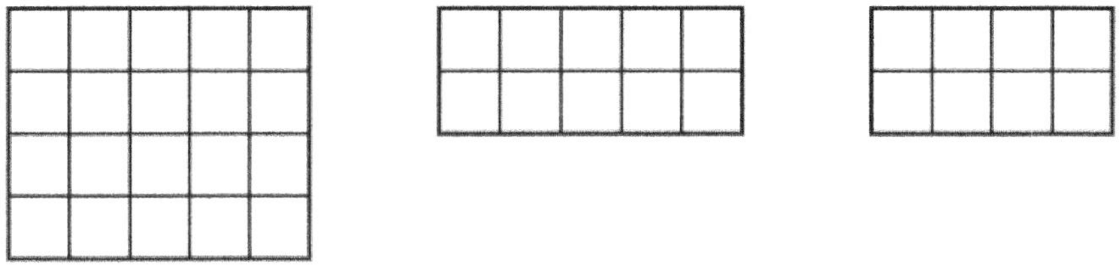

How many unit cubes are in the block?

LET'S UNDERSTAND THE QUESTION TOGETHER

This block is made by joining lots of **equal-sized cubes**.

We are shown **3 faces** of the same solid block.
Each face tells us the size of the block in **2 directions**.

So, these three faces help us figure out the block's **3 side lengths**.

We are not counting just one face.
We need the number of cubes in the **whole solid block**.

WHAT WE KNOW

- ✓ One face is **5 rows and 4 columns**.
- ✓ Onc facc is **5 rows and 2 columns**.
- ✓ One face is **4 rows and 2 columns**.
- ✓ The block is solid.
- ✓ The cubes are all the same size.

WHAT WE NEED TO FIND

We need to find the **total number of unit cubes** in the rectangular block.

Our Plan

1. Write the dimensions of the 3 faces.
2. Find the 3 different side lengths.
3. Multiply those 3 side lengths.

Step-by-Step Solution

Step 1: Write the face dimensions

$\mathbf{5 \times 4}$, $\mathbf{5 \times 2}$, $\mathbf{4 \times 2}$

These are the three faces we can see.

Step 2: Find the three different side lengths

From those faces, the side lengths are: **5, 4, 2**

So, the rectangular block is:

- ✓ 5 cubes in one direction
- ✓ 4 cubes in another direction
- ✓ 2 cubes in the third direction

Step 3: Multiply them

$5 \times 4 \times 2 = 40$

So, the block is made from: 40 unit cubes.

Quick Check

A face with dimensions **5 × 4** has: **20** cubes on that layer.

The block is **2 cubes deep**, so: $\mathbf{20 \times 2 = 40}$

That matches our answer.

Final Answer

The rectangular block is made from **40 unit cubes**.

Common Mistake

A common mistake is to multiply the numbers from just one face, such as:

$5 \times 4 = 20$

That only gives the number of cubes on one layer, not the whole block.
*You must use **all 3 side lengths**.*

Exam Tip

*When you see 3 rectangular faces of a prism, write the face dimensions first. Then look for the **3 different side lengths** and multiply them together.*

Mass and Volume Scaling

QUESTION

Mika has two solid toy bricks made from the same material. The smaller brick has a mass of **480 g**.

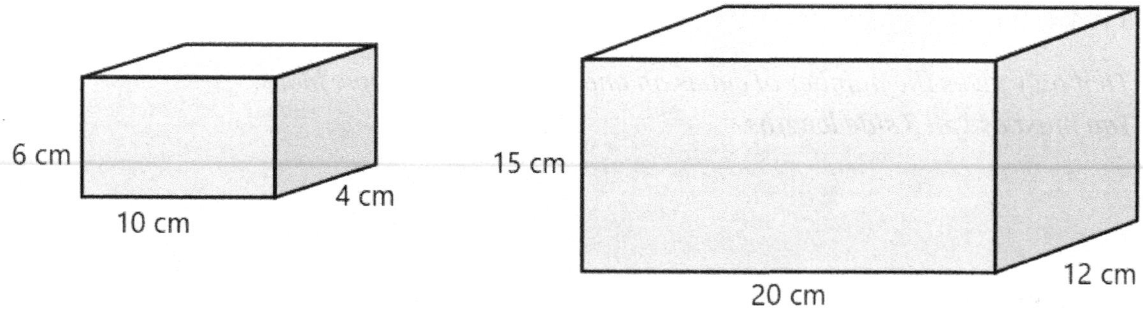

What is the mass of the larger brick, in **kilograms**?

LET'S UNDERSTAND THE QUESTION TOGETHER

We are comparing **two solid bricks** made from the **same material**.

That means:

- ✓ The heavier brick will depend on how much **space** it takes up.
- ✓ If one brick has a larger **volume**, it will also have a larger **mass**.

We already know the mass of the **smaller brick**.

We need to work out how many times bigger the **larger brick** is, then use that to find its mass.

The answer must be given in **kilograms**, not grams.

WHAT WE KNOW

- ✓ Smaller brick dimensions: **4 cm × 6 cm × 10 cm**.
- ✓ Smaller brick mass: **480 g**.
- ✓ Larger brick dimensions: **12 cm × 20 cm × 15 cm**.
- ✓ Both bricks are made from the **same material.**

WHAT WE NEED TO FIND

We need to find the mass of the **larger brick** in **kilograms**.

Our Plan

1. First, find the volume of each brick.
2. Then compare the volumes to see how many times larger the bigger brick is.
3. Finally, multiply the smaller mass by that scale factor and convert the answer to kilograms.

Step-by-Step Solution

Step 1: Find the volume of the smaller brick.

$4 \times 6 \times 10 = 240$

The volume of the smaller brick is: $\mathbf{240\ cm^3}$

Step 2: Find the volume of the larger brick.

$12 \times 20 \times 15 = 3600$

The volume of the larger brick is: $\mathbf{3600\ cm^3}$

Step 3: Compare the two volumes.

$3600 \div 240 = 15$

The larger brick has **15 times** the volume of the smaller brick.

Step 4: Find the mass of the larger brick in grams.

$480 \times 15 = 7200$

So, the larger brick has a mass of: $\mathbf{7200\ g}$

Step 5: Convert grams to kilograms.

$7200\ \mathrm{g} = \mathbf{7.2\ kg}$

Quick Check

The larger brick is **15 times** the volume of the smaller one, so its mass should also be **15 times** as much. And below makes sense:

$480 \times 15 = 7200$ g
7200 g $= 7.2$ kg

Final Answer

The larger brick has a mass of **7.2 kilograms**.

Common Mistake

A common mistake is to compare only one side length instead of the ***whole volume****. For solid shapes, we must multiply* ***length × width × height****.*

Exam Tip

When two solid objects are made from the ***same material****, their masses change in the same ratio as their* ***volumes****. Always find and compare the volumes first.*

Interior Area of a Rectangular Border

QUESTION

Mika is making a garden feature using **8 identical paving stones**.
Each paving stone is a rectangle with length **9 m** and width **3 m**.

What is the area of the shaded grass section?

LET'S UNDERSTAND THE QUESTION TOGETHER

We have **8 identical rectangles**.

Each one measures **9 m by 3 m**.

They are arranged to make a **border** around the outside.

The middle shaded part is a **rectangle**.

We are not finding the area of all the paving stones.

We only want the area of the **inside grass section**.

So, we need to find:

- ✓ The **length** of the inside rectangle.
- ✓ The **width** of the inside rectangle.

Then we can work out its area.

WHAT WE KNOW

- ✓ Each paving stone is **9 m long**.
- ✓ Each paving stone is **3 m wide**.
- ✓ The middle shaded region is a **rectangle**.
- ✓ Area of a rectangle = **length × width.**

WHAT WE NEED TO FIND

The **area of the shaded grass section** in the middle.

Our Plan

- ✓ First, find the **outer width** and **outer height** of the whole design.
- ✓ Then subtract the border thickness from each side to get the inside rectangle.
- ✓ Finally, multiply the inside length and width.

Step-by-Step Solution

Step 1: Find the outer width of the whole design.

Across the top, there are **3 horizontal paving stones**.

Each one is **9 m** long.

So, the total width across the top is:

$$3 \times 9 = 27 \text{ m}$$

There is also **1 vertical paving stone on the right**, and its width is **3 m**.

So, the full outer width is:

$$27 + 3 = 30 \text{ m}$$

So, the **outer width is 30 m**.

Step 2: Find the outer height of the whole design.

A vertical paving stone has height **9 m**.

There is also a top strip that is **3 m** thick.

So, the total outer height is:

$$9 + 3 = 12 \text{ m}$$

So, the **outer height is 12 m.**

Step 3: Find the width of the shaded grass section.

The full outer width is **30 m.**

The border takes up:

- ✓ **3 m on the left.**
- ✓ **3 m on the right.**

So, subtract both sides:

$$30 - 3 - 3 = 24 \text{ m}$$

So, the **inside width is 24 m.**

Step 4: Find the height of the shaded grass section.

The full outer height is **12 m.**

The border takes up:

- ✓ **3 m on the top.**
- ✓ **3 m on the bottom.**

So, subtract both strips:

$$12 - 3 - 3 = 6 \text{ m}$$

So, the **inside height is 6 m.**

Step 5: Find the area of the shaded grass section.

Area of a rectangle:

Area = length × width

So:

$$24 \times 6 = 144$$

So, the area is: 144 m^2.

Quick Check

Let's check the inside dimensions:

- ✓ Width = **24 m**.
- ✓ Height = **6 m**.

Now multiply:

$$24 \times 6 = 144$$

That is sensible for a rectangle inside a larger border.

Final Answer

The area of the shaded grass section is **144 m^2**.

📌 Common Mistake

A common mistake is subtracting the border thickness only ***once*** *instead of from* ***both sides****. Remember: left and right both matter, and top and bottom both matter too.*

💡 Exam Tip

When you see a frame or border made from equal rectangles, it is often quickest to find the ***outside rectangle first****, then subtract the border thickness to get the inside dimensions.*

Packing Trapeziums in a Square

QUESTION

Nai has a **square sheet of card** with side length **24 cm**.

She cuts out identical **right-angled trapeziums**.

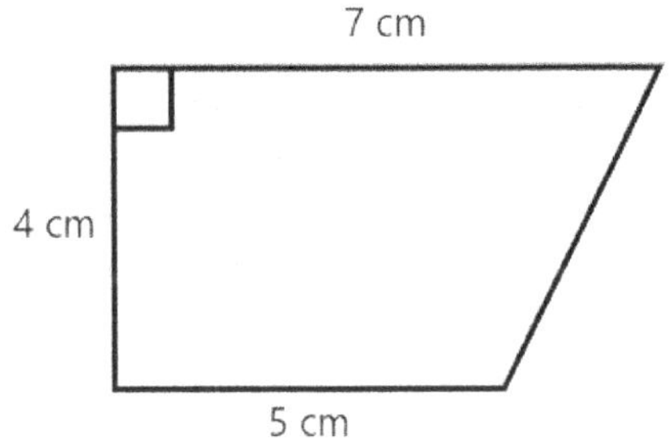

What is the **greatest possible number** of these right-angled trapeziums that Nai can cut from the card?

LET'S UNDERSTAND THE QUESTION TOGETHER

The large piece of card is a **square**, so it measures **24 cm by 24 cm**.

Each small piece is a **right-angled trapezium**. We are told its **height** and the lengths of its **two parallel sides**.

The question wants the **greatest number of complete trapeziums** that can be cut from the square card.

So, we need to compare the size of:

- ✓ The **whole square card**.
- ✓ **One trapezium.**

WHAT WE KNOW

- ✓ The card is a square with side length **24 cm**.
- ✓ The area of the square is **24 × 24**.
- ✓ Each small shape is a **right-angled trapezium**.
- ✓ The trapezium has:
  - height = **4 cm**.
  - parallel sides = **5 cm** and **7 cm**.
- ✓ Area of a trapezium = **½ × (sum of parallel sides) × height**

WHAT WE NEED TO FIND

We need to find the **greatest number of complete right-angled trapeziums** Nai can cut from the square card.

Our Plan

- ✓ First, find the **area of the square card.**
- ✓ Next, find the **area of one trapezium.**
- ✓ Then divide the total area by the area of one trapezium.

Step-by-Step Solution

Step 1: Find the area of the square card.

$$24 \times 24 = 576$$

So, the area of the square card is: **576 cm^2**

Step 2: Add the lengths of the parallel sides of the trapezium.

$$5 + 7 = 12$$

So, the sum of the parallel sides is:

12 cm

Step 3: Use the trapezium area formula.

$$\text{Area of trapezium} = \frac{1}{2} \times (\text{sum of parallel sides}) \times \text{height}$$

Substitute the numbers:

$$\text{Area} = \frac{1}{2} \times 12 \times 4$$

$$= \mathbf{24\ cm^2}$$

Step 5: Divide the total area of the card by the area of one trapezium.

$576 \div 24 = 24$

So, the greatest possible number of trapeziums is: **24**

Quick Check

Let's check by multiplying: $24 \times 24 = 576$

The 24 trapeziums use exactly **576 cm²**, which matches the area of the square card.

Final Answer

Nai can cut out **24 right-angled trapeziums** from the card.

Common Mistake

A common mistake is to forget the $\frac{1}{2}$ *in the trapezium area formula.*

For example, doing: $12 \times 4 = 48$

That would make each trapezium seem too large and give the wrong answer.

Exam Tip

When a question asks for the ***greatest number of shapes*** *that can be cut from a larger shape, find the* ***area of one small shape*** *first, then divide the* ***total area*** *by that amount. Always check that your answer is a whole number of* ***complete shapes****.*

Liquid Volume and Mass Relationship

QUESTION

The image shows the **amount of juice** and the **total weight** of the bottle in two situations: before and after some juice is poured out.

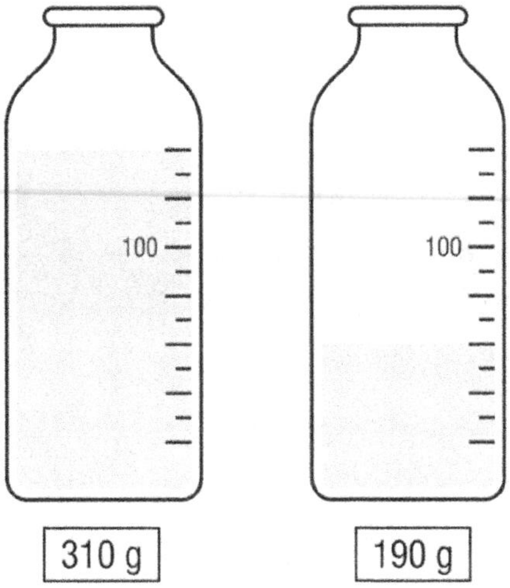

What is the weight of the empty bottle?

LET'S UNDERSTAND THE QUESTION TOGETHER

We are looking at the **same bottle** two times.

The bottle is weighed:

- ✓ Once when it has **140 mL** of juice.
- ✓ Again, when it has **60 mL** of juice.

The **total weight changes** because some juice is poured out.

The bottle itself does **not** change. Only the amount of juice changes.

We need to find the weight of the **empty bottle only**, with **no juice inside**.

WHAT WE KNOW

- ✓ The bottle + **140 mL** of juice weighs **310 g**.
- ✓ The bottle + **60 mL** of juice weighs **190 g**.
- ✓ The amount of juice removed is **140 mL - 60 mL**.
- ✓ The weight lost is **310 g - 190 g.**

WHAT WE NEED TO FIND

We need to find the weight of the **empty bottle**.

Our Plan

First, find how much juice was poured out and how much that juice weighed.

Then work out the weight of **1 mL** of juice.

After that, find the weight of the juice still left in the bottle.

Finally, subtract that juice weight from the total weight to get the weight of the **empty bottle**.

Step-by-Step Solution

Step 1: Find how much weight was lost.

The total weight went from **310 g** to **190 g**.

$310 - 190 = 120$

So, **120 g** of juice was poured out.

Step 2: Find how much juice was poured out.

The juice level went from **140 mL** to **60 mL**.

$140 - 60 = 80$

So, **80 mL** of juice was poured out.

Step 3: Find the weight of 1 mL of juice.

We know **80 mL** of juice weighs **120 g**.

$120 \div 80 = 1.5$

So, **1 mL of juice weighs 1.5 g.**

Step 4: Find the weight of the juice left in the bottle.

There are **60 mL** left.

$60 \times 1.5 = 90$

So, the juice left in the bottle weighs **90 g.**

Step 5: Find the weight of the empty bottle.

The bottle and remaining juice together weigh **190 g.**

So:

$190 - 90 = 100$

So, the empty bottle weighs **100 g.**

Quick Check

If the empty bottle weighs **100 g**, then **140 mL** of juice should weigh:

$140 \times 1.5 = 210$

Then total weight would be:

$100 + 210 = 310$

That matches the question, so our answer is correct.

Final Answer

The empty bottle weighs **100 g.**

Common Mistake

*A common mistake is to subtract the two total weights and think that answer is the bottle's weight. But **120 g** is the weight of the **juice poured out**, not the empty bottle.*

Exam Tip

*When a container is weighed twice, look at **what changed** and **what stayed the same**. The container stays the same, so comparing the two weighing helps you find the weight of the liquid first.*

Logic & Problem Solving

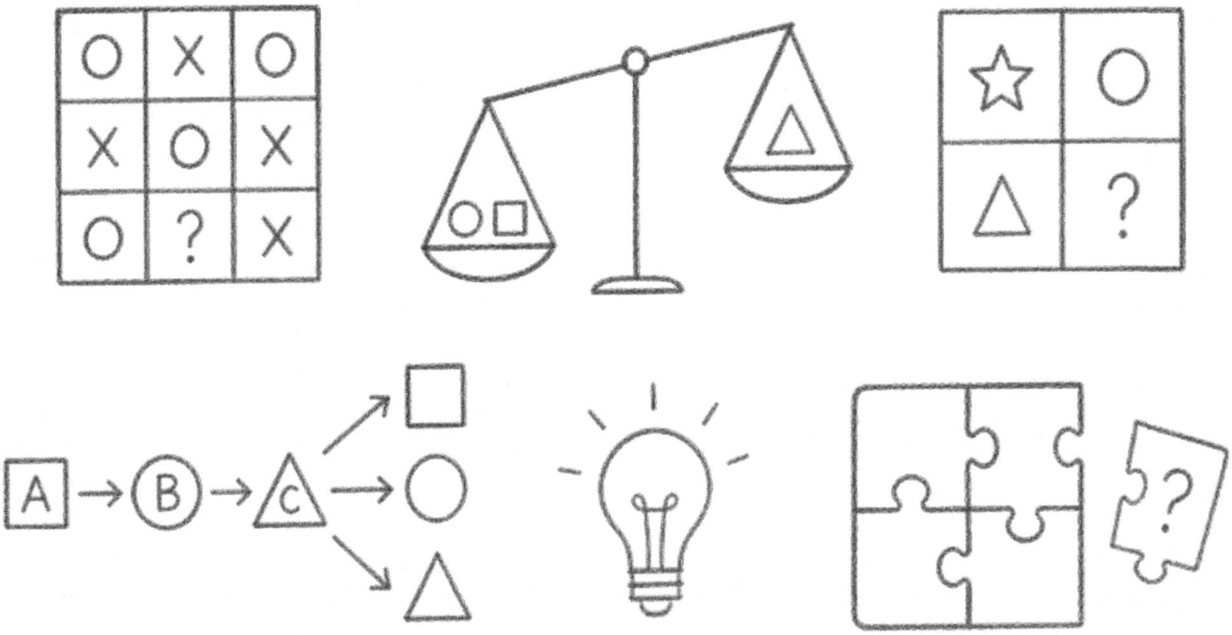

Equivalent Equations with Number Selection

QUESTION

Mika is filling in four empty boxes to make both number sentences true:

$\square - 4 = \square - 1$

$\square \div 2 = \square \div 8$

She must use **four** of these numbers, **once each**:

3, 6, 9, 12, 16

One number will not be used.

Which number is left over?

LET'S UNDERSTAND THE QUESTION TOGETHER

We have **two separate number sentences.**

Each empty box must be filled with a **different number** from the list.

That means we must choose 4 different numbers from the list, using each at most once.

The first sentence is about **subtraction.**

The second sentence is about **division.**

We need to work out which four numbers fit correctly, and then see which number is **left over.**

WHAT WE KNOW

- ✓ The numbers to choose from are:
  - **3, 6, 9, 12, 16**
- ✓ We must use **four numbers once each**
- ✓ First equation: $\square - 4 = \square - 1$
- ✓ Second equation: $\square \div 2 = \square \div 8$
- ✓ One number is **not used**

WHAT WE NEED TO FIND

We need to find **the one number that is not used.**

Our Plan

First, solve the division sentence, since it gives fewer possible pairs.

Then, use the remaining numbers to solve the subtraction sentence.

At the end, check which number is left over.

Step-by-Step Solution

Step 1: Solve the subtraction sentence

We want: $\square - 4 = \square - 1$

Let the left box be a and the right box be b.

So: $a - 4 = b - 1$

Add 4 to both sides: $a = b + 3$

So, the left number must be **3 more** than the right number.

Step 2: Find two numbers from the list with a difference of 3

From **3, 6, 9, 12, 16**, the pairs with a difference of 3 are:

- ✓ 6 and 3
- ✓ 9 and 6
- ✓ 12 and 9

So, the subtraction sentence must use one of these pairs, with the larger number on the left.

Step 3: Solve the division sentence

We want: $\square \div 2 = \square \div 8$

Let the left box be $c$and the right box be d.

So: $c \div 2 = d \div 8$

$$\frac{c}{2} = \frac{d}{8}$$

Multiply both sides by 8:

4c = d

So, the right number must be **4 times** the left number.

Step 4: Find two numbers from the list where one is 4 times the other

From the list:

- $3 \times 4 = 12$ ✓
- $6 \times 4 = 24$ ✗
- $9 \times 4 = 36$ ✗

So, the only possible pair is:

✓ Left box = **3**

✓ Right box = **12**

Check:

$$3 \div 2 = 1.5$$

$$12 \div 8 = 1.5$$

Yes, it works.

Step 5: See which numbers are left for the subtraction sentence

We have used **3** and **12**.

The remaining numbers are: **6, 9, 16**

We need a pair where the left number is **3 more** than the right number.

From these, the pair is: **9 and 6**

Check:

$$9 - 4 = 5$$

$$6 - 1 = 5$$

Yes, it works.

Step 6: Find the unused number

We used: **9 ,6, 3, 12**

The only number not used is: **16**

Quick Check

Let's test both completed sentences:

$9 - 4 = 6 - 1$
$5 = 5$ ✓

$3 \div 2 = 12 \div 8$
$1.5 = 1.5$ ✓

Everything works, so the leftover number is correct.

Final Answer

The number that is not used is **16**.

📌 Common Mistake

*A common mistake is to find a pair that works for **one** equation, but forget to check whether the **remaining numbers** can still make the second equation true.*

💡 Exam Tip

*Start with the **multiplication sentence** first. It gives fewer possible pairs, so it helps you narrow the choices quickly.*

Minimum Button Presses (Integers)

QUESTION

Nai is using a number tool. The number on the screen starts at **18**.

There are two buttons:

- ✓ One button adds **7**.
- ✓ One button subtracts **4**.

Nai can press the buttons in any order. Nai wants the number to become **1**.

What is the **smallest number of button presses** needed?

LET'S UNDERSTAND THE QUESTION TOGETHER

We begin at **18**.

Each button press changes the number:

- ✓ It either goes **up by 7**.
- ✓ Or goes **down by 4**.

We must reach **exactly 1**, not close to it.

We also want to do this using the **fewest total presses**.

WHAT WE KNOW

- ✓ Start = **18**
- ✓ Add button = **+7**
- ✓ Subtract button = **−4**
- ✓ Target = **1**
- ✓ Total presses = $\mathbf{x + y}$

WHAT WE NEED TO FIND

The **smallest total number of presses** that changes **18 into 1**.

Our Plan

We will form an equation and solve it.

Let:

- ✓ x = number of times we press **+7**
- ✓ y = number of times we press **−4**

Step-by-Step Solution

Step 1: Write the equation

$18 + 7x - 4y = 1$

Step 2: Rearrange the equation

Subtract 18 from both sides:

$$7x - 4y = -17$$

Multiply both sides by -1 to make it easier:

$$4y - 7x = 17$$

Step 3: Substitute small values of $\boldsymbol{x}$ into $4y - 7x = 17$.

We can try small values of x.

Try $\boldsymbol{x = 0}$

$$4y - 7(0) = 17$$
$$4y = 17$$
$$y = \frac{17}{4}$$

This is **not** a whole number, so it does not work.

Try $x = 1$

$$4y - 7(1) = 17$$
$$4y - 7 = 17$$

Add 7 to both sides:

$$4y = 24$$
$$y = 6$$

This works.

Step 4: Count the total button presses.

$$x + y = 1 + 6 = 7$$

So, this gives **7 presses**.

Step 5: Check that it is the smallest.

We tested values of x starting from 0 and increasing.

The first valid solution occurs at $x = 1$, giving $y = 6$.

Since larger values of x would increase the total $x + y$, this gives the minimum number of presses.

Quick Check

Start at **18**

Press **+7 once**:

$$18 + 7 = 25$$

Press **−4 six times**:

$$25 - 24 = 1$$

Correct final number

Final Answer

The smallest number of button presses needed is **7**.

Common Mistake

Students sometimes stop when they find a working solution, but forget to check if it is the ***smallest number of presses.***

Exam Tip

When you get an equation like

$$4y - 7x = 17$$

test ***small values first*** *(like* $x = 0,1,2$*). This helps you find the answer quickly without guessing randomly.*

Odd and Even Reasoning

QUESTION

Mika says:

- ✓ A number with a ★ added to a number with a ⬡ gives an **odd** total.
- ✓ A number with a ⬡ multiplied by a number with a ♥ gives an **odd** answer.

Which statements are correct?

1. **★ + ⬡ + ♥ is odd**
2. **★ multiplied by the sum of ⬡ and ♥ is odd**
3. **⬡ multiplied by the sum of ♥ and ★ is odd**

Write the number of each correct statement.

LET'S UNDERSTAND THE QUESTION TOGETHER

Each symbol stands for a whole number.

We are given two clues. One clue is about **addition**. The other clue is about **multiplication**.

These clues tell us whether the numbers must be **odd** or **even**.

Then we check each of the three statements to see which one must be true.

We do **not** need to know the exact numbers. We only need to track whether each symbol is **odd** or **even**.

WHAT WE KNOW

- ✓ **★ + ⬡ is odd**
- ✓ **⬡ × ♥ is odd**

WHAT WE NEED TO FIND

Which of statements **1, 2, and 3** are correct.

Helpful Rules

Addition	Multiplication
✓ even + even = even	✓ even × even = even
✓ even + odd = odd	✓ even × odd = even
✓ odd + even = odd	✓ odd × even = even
✓ odd + odd = even	✓ odd × odd = odd

Our Plan

First, work out whether each symbol must be **odd** or **even**.

Then use the odd-and-even rules to test each statement one at a time.

Step-by-Step Solution

Step 1: Use the multiplication clue

We know:

⬡ × ♥ is odd

From the multiplication rules, the only way to get an **odd** product is:

odd × odd = odd

So, both ⬡ and ♥ are odd.

Step 2: Use the addition clue

We know: **★ + ⬡ is odd**

We already know ⬡ **is odd**.

From the addition rules:

- ✓ **even + odd = odd**
- ✓ **odd + odd = even**

So ★ must be **even**.

Now we know:

- ✓ **★ = even**
- ✓ **⬡ = odd**
- ✓ **♥ = odd**

Step 3: Test statement 1

Statement 1 says:

★ + ⬡ + ♥ is odd

Substitute the types:

even + odd + odd

First:

even + odd = odd

Then:

odd + odd = even

So, statement 1 is **not correct**.

Step 4: Test statement 2

Statement 2 says:

★ multiplied by the sum of ⬡ and ♥ is odd

Substitute the types:

even × (odd + odd)

First simplify inside the brackets:

odd + odd = even

So, we get:

even × even

Now use the multiplication rule:

even × even = even

So, statement 2 is **not correct**.

Step 5: Test statement 3

Statement 3 says:

⬡ multiplied by the sum of ♥ and ★ is odd

Substitute the types:

$$\text{odd} \times (\text{odd} + \text{even})$$

First simplify inside the brackets:

$$\text{odd} + \text{even} = \text{odd}$$

So, we get:

$$\text{odd} \times \text{odd}$$

Now use the multiplication rule:

$$\text{odd} \times \text{odd} = \text{odd}$$

So, statement 3 **is correct**.

Quick Check

Let's test with simple numbers that match the clues:

- ✓ ★ = 2
- ✓ ⬡ = 3
- ✓ ♥ = 5

Check the clues:

- ✓ $2 + 3 = 5$, which is odd.
- ✓ $3 \times 5 = 15$, which is odd.

Now test the statements:

1. $2 + 3 + 5 = 10$, even.
2. $2 \times (3 + 5) = 16$, even.
3. $3 \times (5 + 2) = 21$, odd.

This confirms that only statement **3** works.

Final Answer

The correct statement is 3.

Common Mistake

*A common mistake is forgetting that an **odd** multiplication answer can only happen when **both numbers are odd**.*

*Another common mistake is not simplifying the **brackets first** before multiplying.*

Exam Tip

*First, work out whether each symbol is **odd** or **even**.*

*Then use the **odd/even rules** to test each statement carefully, one step at a time.*

Combinations and Sum Properties

QUESTION

Mika picks **three different number cards** from the set **2, 4, 7, 9**.

Mika adds the three chosen numbers to make one score.

Decide which statement numbers must be written as correct.

1. It is possible for the score to be an **even number**.
2. It is possible for the score to be **divisible by 3**.
3. Every possible score is **more than 15**.

LET'S UNDERSTAND THE QUESTION TOGETHER

Mika is choosing **3 cards out of 4**.

So, each time, **one card is not used**.

The score depends on which 3 cards are chosen.

We need to test the statements against the **possible scores**.

A statement with **"it is possible"** only needs **one example**.

A statement with **"every possible"** must work for **all cases**.

WHAT WE KNOW

1. The number cards are **2, 4, 7, 9**
2. Mika chooses **3 different cards**.
3. The 3 chosen numbers are added to make a score.
4. We must decide which statements are true.

WHAT WE NEED TO FIND

We need to find **which statement numbers are correct**.

Our Plan

1. We will add all 4 numbers first.
2. Then we will subtract the card left out each time to find every possible score.
3. After that, we will test each statement.

Step-by-Step Solution

Step 1: Add all 4 numbers.

$2 + 4 + 7 + 9 = 22$

So, the total of all 4 cards is **22**.

Step 2: Find each possible score by leaving out one card.

If Mika chooses 3 cards, that is the same as taking **22** and subtracting the card not chosen.

✓ Leave out 2: $22 - 2 = 20$

✓ Leave out 4: $22 - 4 = 18$

✓ Leave out 7: $22 - 7 = 15$

✓ Leave out 9: $22 - 9 = 13$

So, the possible scores are: **20, 18, 15, 13**

Step 3: Check Statement 1.

Statement 1 says the score can be **even**.

From the scores, **20** and **18** are even.

So, **Statement 1 is correct**.

Step 4: Check Statement 2.

Statement 2 says the score can be **divisible by 3**.

From the scores, **18** and **15** are divisible by 3.

So, **Statement 2 is correct**.

Step 5: Check Statement 3.

Statement 3 says **every possible score** is more than 15.

But **15** is not more than 15, and **13** is also not more than 15.

So, **Statement 3 is not correct**.

Quick Check

We checked all 4 possible scores, because each one comes from leaving out a different card once.

- ✓ Even score? Yes
- ✓ Divisible by 3? Yes
- ✓ Always more than 15? No

So, our answer is consistent.

Final Answer

The correct statement numbers are **1 and 2**.

Common Mistake

A common mistake is treating "it is possible" and "every possible" as if they mean the same thing.

Exam Tip

*Another valid method is to **list all possible groups of 3 numbers:***

- ✓ *2, 4, 7*
- ✓ *2, 4, 9*
- ✓ *2, 7, 9*
- ✓ *4, 7, 9*

Then add each group and test the statements.

*The first method is usually quicker, but the second method is a great way to make sure you have checked **every possible case**.*

Testing Statements About Totals

QUESTION

At the school fair, **Mika** sold only two kinds of fruit cups: **apple** and **pear**.

Mika sold **41 fruit cups** altogether.

These are three things someone said about the sales:

1. Mika sold **9 more apple cups than pear cups**.
2. Mika sold **3 times as many apple cups as pear cups**.
3. Mika sold **7 more pear cups than apple cups**.

One of these statements cannot be true.

LET'S UNDERSTAND THE QUESTION TOGETHER

We are told there are only **two types** of fruit cups: **apple** and **pear**.

The total number sold is **41**.

Each statement gives a different relationship between the number of apple cups and pear cups.

Our job is to work out which statement is **impossible** when the total is **41**.

So, we are not finding all possibilities. We are checking each statement against the total.

WHAT WE KNOW

Total fruit cups sold = **41**

There were only:

- ✓ **Apple cups.**
- ✓ **Pear cups.**

We need to test these statements:

1. apple = pear + 9
2. apple = 3 × pear
3. pear = apple + 7

WHAT WE NEED TO FIND

We need to find **which statement cannot be correct.**

Our Plan

Test each statement one at a time.

For each one, see whether the two numbers can add up to **41**.

If a statement gives a whole-number answer that totals 41, it could be true.

If it does not, then it cannot be true.

Step-by-Step Solution

Step 1: Test statement 1:

pear + (pear + 9) = 41
2pear + 9 = 41
2pear = 32
pear = 16

So, **apple** = 16 + 9 = 25
So statement 1 can be **correct.**

Step 2: Test statement 2:

pear + 3pear = 41
4pear = 41
pear = 10.25

Not a whole number, so this does **not** work.

Step 3: Test statement 3:

apple + (apple + 7) = 41
2apple = 34
apple = 17, pear = 24

This **works.**

Quick Check

Statements 1 and 3 give whole numbers that add to **41**.

Statement 2 gives **10.25**, so it is impossible.

Final Answer

Statement 2 cannot be correct.

Mika cannot sell part of a fruit cup, so the numbers must be whole numbers. Statement 2 does not give whole numbers that add to 41.

Therefore: **2. Mika sold 3 times as many apple cups as pear cups.**

Common Mistake

A common mistake is checking only whether the statement "sounds possible" instead of testing it with the total.

*Another mistake is forgetting that the number sold must be a **whole number**.*

Exam Tip

*When two amounts add to a total, turn the words into a simple equation and test each statement carefully. Checking whether the answer is a **whole number** is often the key.*

Counting Outfit Combinations

QUESTION

Mika has **4 different scarves** and **4 different jackets**.

Nai has **2 different scarves** and **5 different jackets**.

Each person can make an outfit from their own clothes by choosing **1 scarf** and **1 jacket**.

Who has a larger number of possible outfits, and by how many?

LET'S UNDERSTAND THE QUESTION TOGETHER

Each outfit is made by choosing:

- ✓ **1 scarf**
- ✓ **1 jacket**

Mika chooses from **Mika's own clothes** only.
Nai chooses from **Nai's own clothes** only.

We need to find the total number of outfits for each person.
Then we compare those totals to see **who has more** and **how many more**.

WHAT WE KNOW

- ✓ Mika has **4 scarves**.
- ✓ Mika has **4 jackets**.
- ✓ Nai has **2 scarves**.
- ✓ Nai has **5 jackets**.
- ✓ One outfit = **1 scarf + 1 jacket.**

WHAT WE NEED TO FIND

We need to find which person can make more outfits and the difference between their totals.

Our Plan

We will multiply the number of scarves by the number of jackets for each person. Then we will compare the two answers.

Step-by-Step Solution

Step 1: Find how many outfits Mika can make.

Mika has 4 choices for scarves and 4 choices for jackets.

$4 \times 4 = 16$

So, Mika can make **16 outfits**.

Step 2: Find how many outfits Nai can make.

Nai has 2 choices for scarves and 5 choices for jackets.

$2 \times 5 = 10$

So, Nai can make **10 outfits**.

Step 3: Compare the totals.

- ✓ Mika: **16 outfits**
- ✓ Nai: **10 outfits**

Since 16 is greater than 10, **Mika** has more possible outfits.

Step 4: Find how many more outfits Mika can make.

$16 - 10 = 6$

So, Mika can make **6 more outfits**.

Quick Check

$4 \times 4 = 16$ and $2 \times 5 = 10$.
The difference between 16 and 10 is 6, so the answer makes sense.

Final Answer

Mika has the larger number of possible outfits.

Mika can make **16 outfits**, and Nai can make **10 outfits**.

$$16 - 10 = 6$$

So, **Mika can make 6 more outfits than Nai.**

Common Mistake

A common mistake is to add the numbers of clothes instead of multiplying.
We multiply because ***each scarf can be paired with each jacket.***

Exam Tip

When a question asks for one choice from one group and one choice from another group, ***multiply the numbers of choices first,*** *then compare the totals.*

Distances on a Straight Line

QUESTION

Five bus stops are on one straight road in this order:

Stop P, Stop Q, Stop R, Stop S, Stop T

- Stop P is **360 m** from Stop R.
- Stop P is **900 m** from Stop T.
- Stop Q is **4 times as far from Stop P** as it is from Stop R.
- Stop Q and Stop S are **the same distance from Stop R**.

How far is **Stop S** from **Stop T**?

LET'S UNDERSTAND THE QUESTION TOGETHER

All five bus stops are on a **straight line** and we already know their **order**.

That means:

✓ Stop Q is between **P and R.**

✓ Stop S is between **R and T.**

We are told two full distances starting from **Stop P**:

- From **P to R.**
- From **P to T.**

We are also told how **Q** sits between **P and R.**

Then we are told that **Q and S are equally far from R.**

So, the question is asking for the distance from **S to T.**

WHAT WE KNOW

1. The order is **P, Q, R, S, T**.
2. $PR = 360$ m.
3. $PT = 900$ m.
4. $PQ = 4 \times QR$.
5. $QR = RS$.

WHAT WE NEED TO FIND

We need to find **the distance from Stop S to Stop T**.

Our Plan

First, work out the distances between **P, Q, and R**.

Then use the equal distance fact to find **RS**.

Finally, subtract to find **ST**.

Step-by-Step Solution

Step 1: Look at the distance from P to R

We know:

$$PR = 360 \text{ m}$$

Since Q is between P and R:

$$PQ + QR = 360$$

So, the 360 m is split into **two parts**: $PQ$and QR.

Step 2: Use the clue about Q

We are told:

$PQ = 4 \times QR$

So, the distance from **P to R** is made of:

- ✓ **4 equal parts** for PQ.
- ✓ **1 equal part** for QR.

That means:

$PR = 5$ equal parts

Step 3: Find the size of one equal part

$360 \div 5 = 72$

So, one equal part is **72 m.**

That means:

$QR = 72$ m

Step 4: Find PQ

Since $PQ$is 4 equal parts:

$PQ = 4 \times 72 = 288$ m

Step 5: Find RS

We are told that Q and S are equal distances from R.

So: $RS = QR = 72$ m

Step 6: Find the distance from P to S

To get from P to S, we go from P to R, then from R to S.

$PS = PR + RS$
$PS = 360 + 72 = 432$ m

Step 7: Find ST

We know: $PT = 900$ m

And S is between P and T, so:

$$PT = PS + ST$$
$$900 = 432 + ST$$
$$ST = 900 - 432$$
$$ST = 468 \text{ m}$$

Quick Check

Let's check the important parts:

- ✓ $360 \div 5 = 72$, so $QR = 72$
- ✓ Then $PQ = 4 \times 72 = 288$
- ✓ $288 + 72 = 360$, so that matches PR
- ✓ $RS = 72$, so $PS = 360 + 72 = 432$
- ✓ $900 - 432 = 468$

Everything fits correctly.

Final Answer

Stop S is 468 m from Stop T.

Common Mistake

A common mistake is mixing up the statement:

$PQ = 4 \times QR$

Some students accidentally use $PQ + 4 = QR$ *or* $PR = 4 \times QR$*, but that is not what the question says.*

Exam Tip

When a question says one distance is a certain number of times another, think in ***equal parts****. That often makes the question much easier to solve.*

Distance between points

QUESTION

Mika, Nai and Leo all live on the same straight street.
Mika lives **0.8 km** from Nai.
Leo lives **0.5 km** from Mika.

What are two possible distances between Nai and Leo?

LET'S UNDERSTAND THE QUESTION TOGETHER

All three homes are on **one straight street**. That means the distances are measured along the same line.

We know:

- ✓ The distance from **Mika to Nai**.
- ✓ The distance from **Leo to Mika**.

We need to work out the distance between **Nai and Leo**.

Because the homes are on the same straight street, **Nai and Leo could be on opposite sides of Mika** or on the **same side**. So, there is more than one possible distance.

WHAT WE KNOW

- ✓ Mika is **0.8 km** from Nai.
- ✓ Leo is **0.5 km** from Mika.
- ✓ All three homes are on the **same straight street.**

WHAT WE NEED TO FIND

The distance between **Nai and Leo**.

Our Plan

We will think about the two possible positions on a straight line:

- ✓ Opposite sides of Mika.
- ✓ Same side of Mika.

Then we will calculate the distance in each case.

Step-by-Step Solution

Step 1: Write the two known distances.

Mika to Nai = **0.8 km**
Mika to Leo = **0.5 km**

Step 2: Check the case where Nai and Leo are on **opposite sides** of Mika.

Then the total distance is:

$$0.8 + 0.5 = \mathbf{1.3\ km}$$

So, one possible distance is **1.3 km**.

Step 3: Check the case where Nai and Leo are on the **same side** of Mika.

Then we find the difference:

$$0.8 - 0.5 = \mathbf{0.3\ km}$$

So, the other possible distance is **0.3 km**.

Step 4: State both possible answers.

The distance between Nai and Leo could be:

- ✓ **1.3 km**
- ✓ **0.3 km**

Quick Check

Both answers make sense:

- ✓ If Mika is between them, the distances **add**.
- ✓ If they are on the same side, the distances **subtract.**

Final Answer

Nai and Leo could live either **0.3 km apart** or **1.3 km apart.**

Common Mistake

*A common mistake is to only **add** the distances and forget that the people might live on the **same side** of Mika.*

Exam Tip

*When a question says points are on a **straight line**, always check whether there is **more than one possible arrangement** before choosing your answer.*

Estimating Mass in Tonnes

QUESTION

A **Mercedes G-Wagon** has a mass of **2.4 tonnes**.

Using your own body weight as a guide, about how many **Year 6 students** would have the same mass as the G-Wagon?

LET'S UNDERSTAND THE QUESTION TOGETHER

We are comparing the mass of a **Mercedes G-Wagon** with the mass of **Year 6 students.**

The G-Wagon has a mass of **2.4 tonnes.**

We need to imagine how heavy **one student** is, then work out how many students together would have about the same mass as the car.

This is an **estimate** question, so we are looking for a sensible approximate answer.

WHAT WE KNOW

- ✓ The Mercedes G-Wagon has a mass of **2.4 tonnes.**
- ✓ **1 tonne = 1000 kg.**
- ✓ So, **2.4 tonnes = 2400 kg.**
- ✓ A Year 6 student might weigh about **40 kg.**

WHAT WE NEED TO FIND

We need to find about how many Year 6 students would have a total mass of **2400 kg.**

Our Plan

First, convert tonnes into kilograms.

Then, use a reasonable estimate for one Year 6 student's mass.

Finally, divide the car's mass by the student's mass.

Step-by-Step Solution

Step 1: Convert tonnes to kilograms.

$2.4 \text{ tonnes} = 2400\text{kg}$

Step 2: Estimate the mass of one Year 6 student.

A reasonable estimate is **40 kg**.

Step 3: Divide the total mass by the mass of one student.

$2400 \div 40 = 60$

So about **60 students** would have the same mass.

Step 4: Check that this answer makes sense.

If 60 students each weighed 40 kg, then:

$60 \times 40 = 2400\text{kg}$

That matches the mass of the G-Wagon.

Quick Check

A car that weighs **2400 kg** is much heavier than just a few students.

Since each student is about **40 kg**, needing around **60 students** is reasonable.

Final Answer

About **60 Year 6 students** would have the same mass as the Mercedes G-Wagon.

Common Mistake

*A common mistake is forgetting to change **tonnes into kilograms** before dividing.*

*Another common mistake is using an unrealistic student mass, such as **4 kg** instead of **40 kg**.*

Exam Tip

In estimation questions, pick a sensible weight for one student first.
*Using **40 kg** makes the division fast and helps you avoid calculator-style mistakes.*

Linear Combinations of Scores

QUESTION

At a fair, Mika plays a beanbag tossing game. Points are awarded based on which basket the beanbag lands in

Each successful throw lands in exactly one basket.

1. Explain why a total of **9 points** cannot be scored.
2. What is the **smallest number of successful throws** Mika needs to score exactly **25 points**?

LET'S UNDERSTAND THE QUESTION TOGETHER

Mika can score in only **two ways** each time a beanbag goes into a basket.

A successful throw gives either **7 points** or **4 points**.

That means every total score must be made by **adding 7s and 4s**.

In the first part, we need to see whether **9 points** can be made.

In the second part, we need to reach **25 points** using the **fewest successful throws**.

So, we must think about both the **total score** and the **number of throws**.

WHAT WE KNOW

1. There are **two baskets**.
2. A successful throw into the **small basket** scores **7 points**.
3. A successful throw into the **large basket** scores **4 points**.
4. Part 1 asks about a total of **9 points**.
5. Part 2 asks for the **smallest number of successful throws** to make **25 points.**

WHAT WE NEED TO FIND

1. For part 1, we need to decide whether **9 points** is possible.
2. For part 2, we need to find the **fewest successful throws** needed to score **25 points** exactly.

Our Plan

1. We will test totals by adding **4s** and **7s**.
2. First, we will check whether 9 can be made.
3. Then, for 25, we will try using more **7-point throws** first, because larger scores can help us use fewer throws.

Step-by-Step Solution

Part 1: Can Mika score 9 points?

Step 1: Write the possible scores from one successful throw.

One successful throw can score **4** or **7**.

Step 2: Check totals from two successful throws.

- $4 + 4 = 8$
- $4 + 7 = 11$
- $7 + 7 = 14$

Step 3: Look for 9.

The totals near 9 are **8** and **11**, but **9** does not appear.

Step 4: Decide the answer.

Since Mika can only add 4s and 7s, a total of **9 points cannot be scored.**

Part 2: What is the smallest number of successful throws to score 25 points?

Step 5: Test whether **3 throws** could be enough.

The greatest score from 3 successful throws is:

$7 + 7 + 7 = 21$

So, **3 throws are not enough** to reach 25.

Step 6: Try **4 successful throws.**

Check some totals:

- $7 + 7 + 7 + 7 = 28$
- $7 + 7 + 7 + 4 = 25$

Step 7: Find the exact match.

$7 + 7 + 7 + 4 = 25.$ so, **4 successful throws** work.

Step 8: Check that it is the smallest number.

We already showed that **3 throws cannot reach 25**, so **4** is the smallest possible number.

Quick Check

For part 1, 9 is not possible because adding 4s and 7s gives totals like **8, 11, 14**, not 9.

For part 2, $7 + 7 + 7 + 4 = 25$. So the score is correct, and **4 throws** is the minimum.

Final Answer

A total of **9 points cannot be scored**, and the **smallest number of successful throws needed to score 25 points is 4**.

Common Mistake

*A common mistake is to find a combination that makes **25** but not check whether it uses the **fewest successful throws**.*

Exam Tip

*When a question asks for the **smallest number**, try the **larger score first**. This helps you reach the target using fewer throws.*

Patterns & Sequences

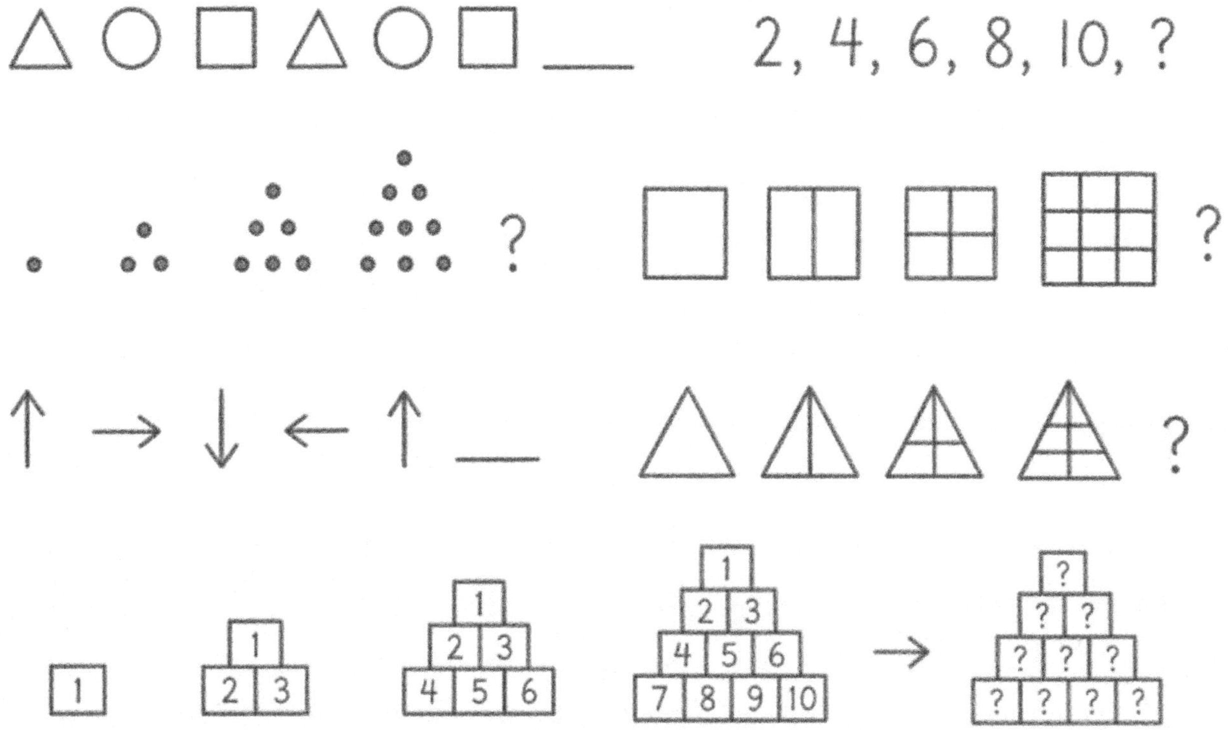

Multiplicative Number Patterns

QUESTION

Mika is writing a number pattern that follows the same rule each time:

What number should go in the empty space?

LET'S UNDERSTAND THE QUESTION TOGETHER

We are looking at a **number pattern**.

The numbers are going in **order**, and each new number is made using the **same rule**.

One number is missing in the middle.

Our job is to work out what number fits so the pattern stays correct all the way through.

WHAT WE KNOW

1. The pattern starts with **5, 10, 20**.
2. Then there is **one missing number**.
3. After that, the pattern continues with **80, 160**.
4. The rule stays the **same each time.**

WHAT WE NEED TO FIND

We need to find the **missing number** in the pattern.

Our Plan

We will look closely at how the numbers are changing from one term to the next.

Then we will use that same rule to fill in the missing number.

Step-by-Step Solution

Step 1: Compare the first two numbers.

$5 \rightarrow 10$

To get from 5 to 10, we **multiply by 2**.

Step 2: Check the next pair.

$10 \rightarrow 20$

To get from 10 to 20, we also **multiply by 2**.

Step 3: Use the same rule to find the missing number.

$20 \times 2 = 40$

So, the missing number is **40**.

Step 4: Check that the pattern still works after that.

$40 \times 2 = 80$

Yes, that matches the next number.

Step 5: Check one more step.

$80 \times 2 = 160$

That also matches.

So, the rule is correct.

Quick Check

The numbers are doubling each time:

5,10,20,40,80,160

This fits perfectly, so our answer makes sense.

Final Answer

The missing number is **40**.

📌 Common Mistake

*A common mistake is to look at the pattern as if it is **adding** each time instead of checking whether it is **multiplying**. Always test the rule carefully.*

💡 Exam Tip

*When you see a number pattern, check for simple operations first: **add**, **subtract**, **multiply**, or **divide**. Then test the rule on more than one pair of numbers.*

Difference Pyramid Patterns

QUESTION

Nai is using a number machine.

The machine works like this: each new row is made by finding the **positive difference** between each pair of neighbouring numbers in the row below.

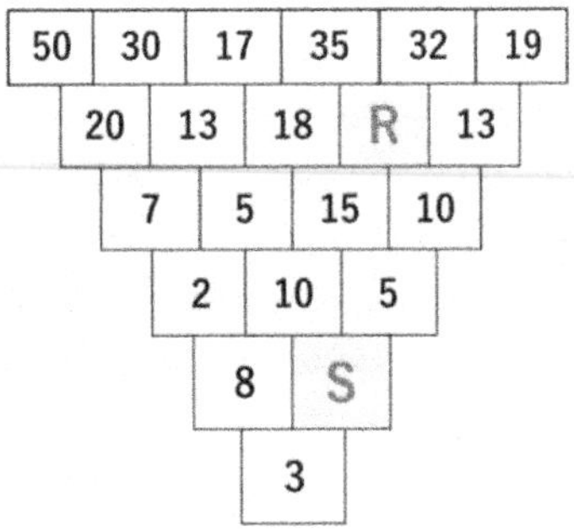

Find the values of **R** and **S**.

LET'S UNDERSTAND THE QUESTION TOGETHER

We start with a row of 6 numbers.

Then a new row is made from the row below it.

Each number in the new row comes from **two neighbouring numbers** underneath.

We are told the rule uses the **positive difference**, which means we subtract the smaller number from the larger number.

Two numbers are missing: **R** and **S**.

We need to work out what those two missing numbers must be so the whole pattern works correctly.

WHAT WE KNOW

- ✓ The first row is: 50, 30, 17, 35, 32, 19
- ✓ Each new row uses the **positive difference** of neighbouring numbers
- ✓ The next rows are:

20, 13, 18, **R**, 13

7, 5, 15, 10

2, 10, 5

8, **S**

3

WHAT WE NEED TO FIND

We need to find R and S.

Our Plan

We will use the pattern rule carefully, one pair at a time.

First, we will find **R** from the first two rows.

Then, we will find **S** from the last two rows.

Finally, we will check that the final number still works.

Step-by-Step Solution

Step 1: Use the first row to make the second row

Find the positive difference between each pair of neighbouring numbers:

- ✓ 50 and 30 → 50 - 30 = 20
- ✓ 30 and 17 → 30 - 17 = 13
- ✓ 17 and 35 → 35 - 17 = 18
- ✓ 35 and 32 → 35 - 32 = 3
- ✓ 32 and 19 → 32 - 19 = 13

So, the second row is: **20, 13, 18, 3, 13**

This means: **R = 3**

Step 2: Check the next row

Now use the row: 20, 13, 18, 3, 13

Find the positive differences:

- ✓ 20 and 13 → 20 - 13 = 7
- ✓ 13 and 18 → 18 - 13 = 5
- ✓ 18 and 3 → 18 - 3 = 15
- ✓ 3 and 13 → 13 - 3 = 10

So, the next row is: **7, 5, 15, 10**

That matches the question.

Step 3: Make the next row

Now use: 7, 5, 15, 10

Find the positive differences:

- ✓ 7 and 5 → 7 - 5 = 2
- ✓ 5 and 15 → 15 - 5 = 10
- ✓ 15 and 10 → 15 - 10 = 5

So, the next row is: **2, 10, 5**

That also matches.

Step 4: Find S

Now use: 2, 10, 5

Find the positive differences:

- ✓ 2 and 10 → 10 - 2 = 8
- ✓ 10 and 5 → 10 - 5 = 5

So, the next row is: **8, 5.** This means: **S = 5**

Step 5: Check the top number

Now use: 8, 5

Find the positive difference: 8 - 5 = 3

The top number is **3**, so the pattern works perfectly.

Quick Check

Let's test both answers:

- ✓ **R = 3** because the difference between 35 and 32 is 3
- ✓ **S = 5** because the difference between 10 and 5 is 5
- ✓ The final check gives 8 - 5 = 3, which matches the top

So, both missing numbers are correct.

Final Answer

R = 3 and S = 5.

📌 Common Mistake

A common mistake is forgetting that the question asks for the ***positive difference****.*

That means you do ***not*** *keep a negative answer. Always subtract the smaller number from the larger number.*

💡 Exam Tip

When a pattern question has missing numbers, write out each new row carefully from left to right. This helps you see exactly where each missing value comes from.

Reverse Number Pattern Rules

QUESTION

Mika writes a number pattern.

She starts with a number and then repeats these two rules in order:

- Add 40.
- Subtract 250.

What is the 1st number?

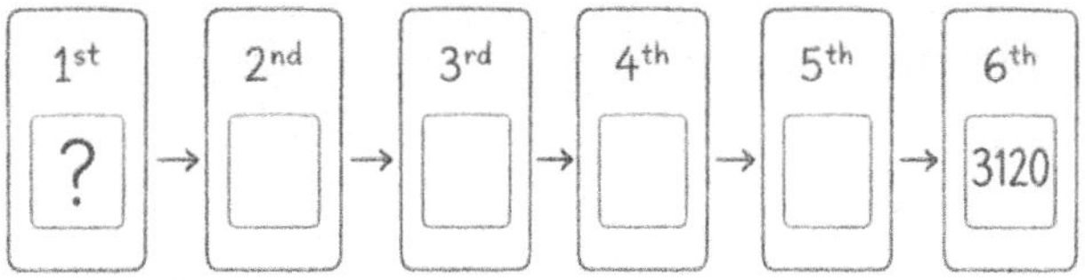

LET'S UNDERSTAND THE QUESTION TOGETHER

Mika begins with a starting number, but we do not know what it is yet.

She keeps following the same **two-step pattern**:

- First **add 40**.
- Then **subtract 250**.

This repeats again and again to make the sequence.

We are told the value of the **6th number**, which is **3120**.

Our job is to find the **starting number**, which is the **1st number**.

WHAT WE KNOW

The 1st number is unknown.

The rules repeat in this order:

✓ **Add 40**

✓ **Subtract 250**

The **6th number is 3120**.

WHAT WE NEED TO FIND

The **1st number** in the sequence.

Our Plan

We will let the 1st number be **x**.

Then we will write each term of the sequence in terms of **x**.

After that, we will use the fact that the 6th number is **3120** to make an equation and solve it.

Step-by-Step Solution

Step 1: Let the 1st number be **x**.

So: 1st number = **x**

Step 2: Find the 2nd number.

The rule says **add 40**.

So: 2nd number = **x + 40**

Step 3: Find the 3rd number.

Now subtract **250** from the 2nd number:

3rd number = **x + 40 - 250**

Simplify: 3rd number = **x – 210**

Step 4: Find the 4th number.

Now add **40**:

4th number = **x - 210 + 40**

Simplify: 4th number = **x - 170**

Step 5: Find the 5th number.

Now subtract **250**:

5th number = **x - 170 - 250**

Simplify: 5th number = **x – 420**

Step 6: Find the 6th number.

Now add **40**:

6th number = **x - 420 + 40**

Simplify: 6th number = **x – 380**

Step 7: Use the information from the question.

We are told the 6th number is **3120**.

So: **x - 380 = 3120**

Add 380 to both sides:

x = 3120 + 380

x = 3500

Quick Check

Let's check by starting at **3500**:

- ✓ 1st: **3500**
- ✓ 2nd: **3500 + 40 = 3540**
- ✓ 3rd: **3540 - 250 = 3290**
- ✓ 4th: **3290 + 40 = 3330**
- ✓ 5th: **3330 - 250 = 3080**
- ✓ 6th: **3080 + 40 = 3120**

The 6th number matches, so the answer is correct.

Final Answer

The 1st number in the pattern is **3500**.

Alternative Method (Working Backwards)

Start from 3120 and reverse each step: subtract 40, add 250, subtract 40, add 250, subtract 40.

This gives 3500.

Common Mistake

*A common mistake is getting the signs mixed up and writing **x + 210** instead of **x - 210** after doing **+40** and then **-250**.*

Exam Tip

*When a sequence starts with an unknown number, call it **x** and write each new term carefully from the one before it. This helps turn the pattern into an equation you can solve.*

Growing Number Sentence Patterns

QUESTION

Mika is looking at a growing pattern of number sentences:

a) Write **Term 15** in this pattern.

b) Nai says that **48 + 36 = 84** is in the same pattern.
Is Nai correct? Explain your answer.

LET'S UNDERSTAND THE QUESTION TOGETHER

We are given a pattern of number sentences.

Each term has:

- ✓ A first number.
- ✓ A second number.
- ✓ A total.

The numbers are growing in a regular way.

In part **a)**, we need to write **Term 15**.

In part **b)**, we need to check whether **48 + 36 = 84** follows the same pattern.

Once we find the rule, we should use it directly instead of doing extra manual work.

WHAT WE KNOW

1. Term 1: $\mathbf{4 + 3 = 7}$
2. Term 2: $\mathbf{8 + 6 = 14}$
3. Term 3: $\mathbf{12 + 9 = 21}$
4. Term 4: $\mathbf{16 + 12 = 28}$

WHAT WE NEED TO FIND

We need to find:

- ✓ The number sentence for **Term 15**.
- ✓ Whether $\mathbf{48 + 36 = 84}$ is in the same pattern.

Our Plan

We will:

1. find the pattern rule.
2. use it directly for **Term 15**.
3. use it directly to test Nai's sentence.

Step-by-Step Solution

Step 1: Look for the pattern.

- ✓ First number: 4, 8, 12, 16, ... so it increases by **4.**
- ✓ Second number: 3, 6, 9, 12, ... so it increases by **3**.
- ✓ Total: 7, 14, 21, 28, ... so it increases by **7**.

So, Term $\boldsymbol{n}$ is:

$$4n + 3n = 7n$$

Step 2: For **Term 15**:

$$4 \times 15 + 3 \times 15 = 7 \times 15$$
$$60 + 45 = 105$$

So, **Term 15 is** $60 + 45 = 105$.

Step 3: Check Nai's sentence:

$48 + 36 = 84$
$48 = 4 \times 12, 36 = 3 \times 12, 84 = 7 \times 12$

So, it matches

$$4n + 3n = 7n$$

when $n = 12$.

So, **Nai is correct**.

Quick Check

For $n = 15$:

$$4(15) + 3(15) = 60 + 45 = 105$$

For $n = 12$:

$$4(12) + 3(12) = 48 + 36 = 84$$

Both fit the pattern.

Final Answer

a) Term 15 is **60 + 45 = 105**.

b) Yes, Nai is correct because **48 + 36 = 84** matches $4n + 3n = 7n$ when $n = 12$.

Common Mistake

A common mistake is continuing term by term even after the rule is clear. That creates extra work and can lead to mistakes.

Exam Tip

When you spot a clear pattern, write it as a rule using n. Then use the rule straight away instead of doing unnecessary manual steps.

Perfect Squares in Growth Patterns

QUESTION

At the school fair, Nai stacks **orange juice boxes** in a growing pattern.

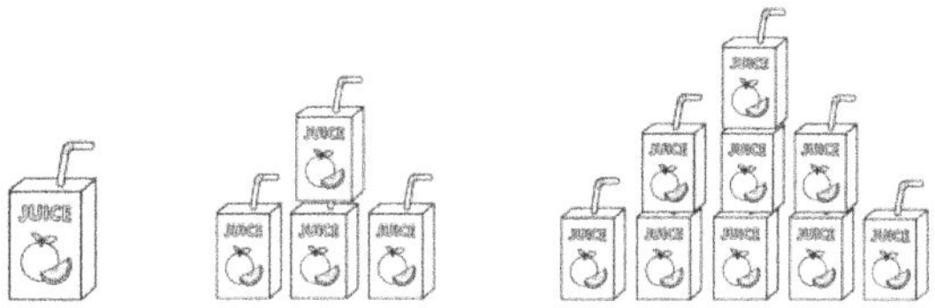

If the same pattern continues:

a) How many orange juice boxes are needed to make a stack **6 juice boxes high**?
b) If a stack uses **49 orange juice boxes**, how many juice boxes high is the stack?

LET'S UNDERSTAND THE QUESTION TOGETHER

This question has **two parts**, but both use the **same pattern**.

The total number of orange juice boxes changes as the **height of the stack** changes.

We are told:

- ✓ Height 1 uses 1 box.
- ✓ Height 2 uses 4 boxes.
- ✓ Height 3 uses 9 boxes.

In part **a**, we know the **height** and need to find the **total number of boxes**.

In part **b**, we know the **total number of boxes** and need to find the **height**.

So, in one part we go **from height to total**, and in the other part we go **from total to height**.

WHAT WE KNOW

1. A **1-high** stack uses **1** orange juice box.
2. A **2-high** stack uses **4** orange juice boxes.
3. A **3-high** stack uses **9** orange juice boxes.
4. The pattern continues in the same way.
5. Part **a** gives the height: **6**
6. Part **b** gives the total number of boxes: **49**

WHAT WE NEED TO FIND

a) The total number of orange juice boxes in a **6-high** stack

b) The height of a stack that uses **49** orange juice boxes

Our Plan

First, find the rule in the pattern. Then use that rule to solve both parts. For the second part, work backwards from the total number of boxes to the height.

Step-by-Step Solution

Step 1: Write the totals in order.

1, 4, 9

Step 2: Look at how each total is made.

✓ $1 = \mathbf{1 \times 1}$

✓ $4 = \mathbf{2 \times 2}$

✓ $9 = \mathbf{3 \times 3}$

So, the pattern is:

total orange juice boxes = height × height

Part a

Step 3: Use the rule for a stack that is 6 boxes high.

$6 \times 6 = \mathbf{36}$

Step 4: Write the answer for part a.

A stack **6 juice boxes high** needs **36 orange juice boxes.**

Part b

Step 5: Work backwards from 49.

We need to find a number that multiplies by itself to make 49.

Step 6: Test the square number.

$7 \times 7 =$ **49**

Step 7: Write the answer for part b.

A stack that uses **49 orange juice boxes** is **7 juice boxes high**.

Quick Check

Check part **a**: $6 \times 6 = 36$, so that works.

Check part **b**: $7 \times 7 = 49$, so the height is 7.

Both answers match the square number pattern.

Final Answer

a) A stack **6 juice boxes high** needs **36 orange juice boxes**.

b) A stack that uses **49 orange juice boxes** is **7 juice boxes high**.

Common Mistake

*A common mistake is to use **addition** instead of recognising **square numbers**.*
*For part **b**, some students may also forget to work backwards and may not check which number squared equals 49.*

Exam Tip

*When a pattern question gives totals like **1, 4, 9**, think of **square numbers**.*
*If the question gives the total and asks for the height, ask yourself: **What number multiplied by itself gives this total?***

Repeating Letter Patterns

QUESTION

A row of letters is written like this:

What are the **92nd** and **93rd** letters, in that order?

LET'S UNDERSTAND THE QUESTION TOGETHER

The letters are written in a fixed order.

We can see the same block of letters keeps going:
A, C, A, D

Then the row continues in that same order.

We only need to find the letters in the **92nd place** and the **93rd place**.

WHAT WE KNOW

The letters go in this order:

- ✓ 1st: A
- ✓ 2nd: C
- ✓ 3rd: A
- ✓ 4th: D

Then the same 4-letter block continues.

We need to find:

- ✓ the **92nd** letter
- ✓ the **93rd** letter

WHAT WE NEED TO FIND

We need to work out which letters are in positions **92** and **93**.

Our Plan

We will use groups of **4 letters**.

Then we will see where 92 and 93 fit inside one 4-letter group.

Step-by-Step Solution

Step 1: Look at one full group.

One full group is:

1. A
2. C
3. A
4. D

Step 2: Find where the 92nd letter lands.

$92 \div 4 = 23$ remainder 0

A remainder of **0** means the 92nd letter is the **4th** letter in the group.

The 4th letter is **D**.

So, the **92nd letter is D**.

Step 3: Find where the 93rd letter lands.

$93 \div 4 = 23$ remainder 1

A remainder of **1** means the 93rd letter is the **1st** letter in the group.

The 1st letter is **A**.

So, the **93rd letter is A**.

Step 4: Write the answer in order.

The 92nd and 93rd letters are **D, A**.

Quick Check

The 92nd position is exactly at the end of a 4-letter group, so it should be the 4th letter, **D**.

The next letter starts the next group, so the 93rd letter should be **A**.

That matches our answer.

Final Answer

The **92nd letter is D** and the **93rd letter is A**.

Common Mistake

*A common mistake is thinking a remainder of **0** means the 1st letter.*
*It actually means the position lands on the **last letter** of the group.*

Exam Tip

When letters or shapes are shown in a continuing row, first find how many items are in one full group. Then use division to locate the position quickly.

Coordinate Patterns on a Grid

QUESTION

Nai is planting trees in a repeating pattern on the grid shown. What is the grid reference of tree 52?

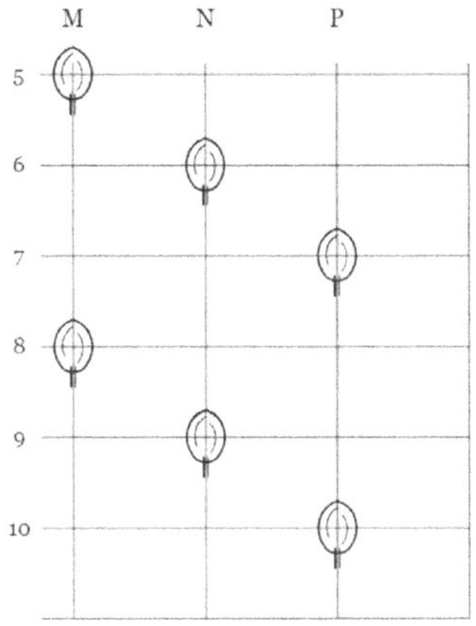

LET'S UNDERSTAND THE QUESTION TOGETHER

This question shows a **pattern** in grid references.

Each tree has a **column letter** and a **row number**.

The column letters repeat in a cycle. The row numbers increase steadily.

We need to find exactly where **tree 52** will be planted.

WHAT WE KNOW

1. Tree 1 is at **M5**.
2. Tree 2 is at **N6**.
3. Tree 3 is at **P7**.
4. Tree 4 is at **M8**.
5. Tree 5 is at **N9**.
6. Tree 6 is at **P10**.
7. The pattern keeps repeating.
8. We need the position of **tree 52.**

WHAT WE NEED TO FIND

We need to find the **grid reference of tree 52**.

Our Plan

1. Look for the pattern in the **column letters**.
2. Look for the pattern in the **row numbers**.
3. Use both patterns to find **tree 52.**

Step-by-Step Solution

Step 1: Look at the column letters.

The letters go:

M, N, P, M, N, P, ...

So, the letters repeat every **3 trees**.

Step 2: Work out which letter tree 52 uses.

Divide 52 by 3:

$52 \div 3 = 17$ remainder 1

A remainder of **1** means we use the **first letter** in the repeating pattern.

The pattern is:

- ✓ 1st letter = **M**
- ✓ 2nd letter = **N**
- ✓ 3rd letter = **P**

So, tree 52 is in column **M**.

Step 3: Look at the row numbers.

The row numbers are: 5, 6, 7, 8, 9, 10, ...

They increase by **1** each time.

Tree **1** is at row **5**, so the row number is always **4 more than the tree number**.

Step 4: Find the row number for tree 52.

$52 + 4 = 56$

So, tree 52 is in row **56**.

Step 5: Put the column and row together.

Column = **M**
Row = **56**

So, the grid reference is **M56**.

Quick Check

Let's check the last few letter positions:

1. Tree 49 gives remainder 1, so **M**.
2. Tree 50 gives remainder 2, so **N**.
3. Tree 51 gives remainder 0, so **P**.
4. Tree 52 goes back to **M**.

The row number also fits because the rows go up by 1 each time, so tree 52 is at row **56**.

Final Answer

The grid reference of **tree 52** is **M56**.

Common Mistake

*A common mistake is to treat remainder **0** as the first letter in the pattern. In a 3-letter cycle, remainder 1 is the first letter, remainder 2 is the second letter, and remainder 0 is the third letter.*

Exam Tip

When a pattern repeats, write out the cycle clearly first. Then use division and the remainder to find the correct position quickly.

Locating Numbers in Rows and Columns

QUESTION

Use the pattern shown in the grid.

1	3	5	7
9	11	13	15
⋮	⋮	⋮	⋮

Mika writes odd whole numbers in rows of 4.

Which row and which column is the number **87** in?

LET'S UNDERSTAND THE QUESTION TOGETHER

Mika is writing only the **odd numbers**.

Each row has **4 numbers**.

That means after every 4 odd numbers, a **new row** begins.

We need to work out exactly where **87** appears:

- ✓ Which **row**.
- ✓ Which **column**.

So, we are finding its position in the pattern, not just checking whether it is odd.

WHAT WE KNOW

1. The numbers are **odd whole numbers**.
2. Each row has **4 numbers**.
3. The pattern keeps going in the same order.
4. We need to find the position of **87**.

WHAT WE NEED TO FIND

We need to find the **row number** and the **column number** for **87**.

Our Plan

First, work out **which odd number 87 is** in the pattern.

Then, use groups of **4** to find the row.

Finally, use the leftover part to find the column.

Step-by-Step Solution

Step 1: Work out how many odd numbers come before 87.

The odd numbers go:

1, 3, 5, 7, ...

To find the position of 87 in the odd-number pattern, use:

$$\frac{87+1}{2} = 44$$

So, **87 is the 44th odd number**.

Step 2: Put the odd numbers into rows of 4.

Each row has 4 numbers, so divide 44 by 4:

$$44 \div 4 = 11$$

This means 44 fits exactly into **11 rows**.

Step 3: Work out the row.

Since the 44th odd number ends exactly at the end of a row, **87 is in Row 11**.

Step 4: Work out the column.

Because 44 divided by 4 has **no remainder**, 87 must be in the **last column** of that row.

So, it is in **Column 4**.

Quick Check

Row 1 has odd numbers 1 to 7.
Each new row adds 4 more odd numbers.

So:

- ✓ Row 10 ends at the 40th odd number.
- ✓ Row 11 ends at the 44th odd number.

Since 87 is the 44th odd number, it must be at the **end of Row 11**, which is **Column 4**.

Final Answer

The number **87 is in Row 11, Column 4**.

📌 Common Mistake

*A common mistake is to divide **87 by 4** straight away.*
*That does not work because the pattern is made from **odd numbers only**, not all whole numbers.*

💡 Exam Tip

*When numbers are arranged in equal rows, first find the number's **position in the pattern**, then use **division** to locate the row and column.*

Common Multiples of Intervals

QUESTION

At a school fair, three effects happen at different time intervals after the fair opens.

- ✓ A **bubble machine** turns on every **4 minutes**.
- ✓ A **music jingle** plays every **6 minutes**.
- ✓ A **confetti popper** goes off every **9 minutes**.

All three happen together exactly at **8:00 am**.

How many times between **8:01 am and 9:59 am** do the **music jingle** and the **confetti popper** happen together **without** the bubble machine turning on at the same time?

LET'S UNDERSTAND THE QUESTION TOGETHER

This question is about **repeating events.**

Each effect happens again and again after a certain number of minutes.

- ✓ The bubble machine repeats every **4 minutes.**
- ✓ The music jingle repeats every **6 minutes.**
- ✓ The confetti popper repeats every **9 minutes.**

At **8:00 am**, all three happen together.

After that, they keep repeating on their own patterns.

We are only interested in times when:

- ✓ The **music jingle** and **confetti popper** happen together.
- ✓ But the **bubble machine does not**.

So, we need to find the times when the **6-minute pattern** and the **9-minute pattern** match, then remove the times when the **4-minute pattern** also matches.

WHAT WE KNOW

- ✓ Bubble machine: every **4 minutes.**
- ✓ Music jingle: every **6 minutes.**
- ✓ Confetti popper: every **9 minutes.**
- ✓ All three happen together at **8:00 am.**
- ✓ We count only from **8:01 am to 9:59 am.**
- ✓ We want times when the **6-minute** and **9-minute** events happen together.
- ✓ We do **not** want times when the **4-minute** event also happens.

WHAT WE NEED TO FIND

We need to find the number of times the **music jingle** and **confetti popper** happen together during the given time period, **without** the bubble machine.

Our Plan

We will use **LCM.**

LCM means **Lowest Common Multiple.**

- ✓ A **multiple** of a number is what you get when you multiply it by 1, 2, 3, 4, and so on.
- ✓ For example, multiples of 6 are: **6, 12, 18, 24, 30...**
- ✓ Multiples of 9 are: **9, 18, 27, 36...**

A **common multiple** is a number that appears in both lists.

The **lowest common multiple** is the **smallest number that both numbers share.**

We use LCM when two or more things are **repeating in cycles** and we want to know **when they will happen together.**

That is exactly what is happening here:

- ✓ The music repeats every **6 minutes.**
- ✓ The confetti repeats every **9 minutes.**

So, we use the **LCM of 6 and 9** to find when they happen together.

Then we also use LCM again to find when the **bubble machine joins in too**, so we can remove those times.

Step-by-Step Solution

Step 1: Find when the music jingle and confetti popper happen together.

We need the **LCM of 6 and 9**.

First, list some multiples:

Multiples of **6:** *6, 12, **18**, 24, 30, 36*

Multiples of **9:** *9, **18**, 27, 36*

The first number that appears in both lists is: **18**

So: $\mathbf{LCM(6, 9) = 18}$

This means the music jingle and confetti popper happen together every **18 minutes**.

Step 2: List all the 18-minute times between 8:01 am and 9:59 am.

Starting from **8:00 am**, the common occurrences happen every **18 minutes**:

- ✓ **8:18**
- ✓ **8:36**
- ✓ **8:54**
- ✓ **9:12**
- ✓ **9:30**
- ✓ **9:48**
- ✓ **10:06**

But the question only goes up to **9:59 am**, so we stop before **10:06**.

So, the music jingle and confetti popper happen together **6 times** in the time period.

Step 3: Find when the bubble machine also happens at the same time.

Now we need to remove any of those times when the **bubble machine** also turns on.

The bubble machine repeats every **4 minutes**.

We already know the jingle and confetti happen together every **18 minutes**.

So now we find: $\mathbf{LCM(18, 4)}$

List some multiples:

Multiples of 1**8:** *18, **36**, 54, 72, 90, 108*
Multiples of **4:** *8, 12, 16, 20, 24, 28, 32, **36***

The first common multiple is: **36.**

So: $\text{LCM}(18, 4) = 36$

This means **all three** happen together every **36 minutes.**

Step 4: List all the 36-minute times between 8:01 am and 9:59 am.

Starting from 8:00 am, add 36 minutes each time: **8:36, 9:12, 9:48, 10:24**

But **10:24** is too late. So, all three happen together **3 times** in the time period.

Step 5: Subtract the times we do not want.

The music jingle and confetti popper happen together **6 times.**

But **3** of those times also include the bubble machine.

So, we subtract: $\mathbf{6 - 3 = 3}$

Quick Check

The jingle and confetti match every **18 minutes**, so over nearly 2 hours, **6 matches** is reasonable.

The bubble machine joins them every **36 minutes**, which is every second 18-minute match.

So, removing **3** from **6** makes sense.

Final Answer

The music jingle and the confetti popper happen together **3 times** without the bubble machine turning on.

Common Mistake

*A common mistake is using LCM once, finding **6** times, and stopping there.*

*But the question says **without** the bubble machine, so we must also remove the times when **all three** happen together.*

Exam Tip

*Use **LCM** whenever events repeat in equal cycles and you need to know **when they line up together**. In time questions like this, LCM helps you find the repeating meeting point of the patterns.*

Net Movement on a Grid

QUESTION

A treasure map shows a grid. Mika starts on square **C4**. She follows these treasure-map instructions in a repeating pattern:

north, east, south, west

This image shows only a partial view of the map, specifically the bottom corner.

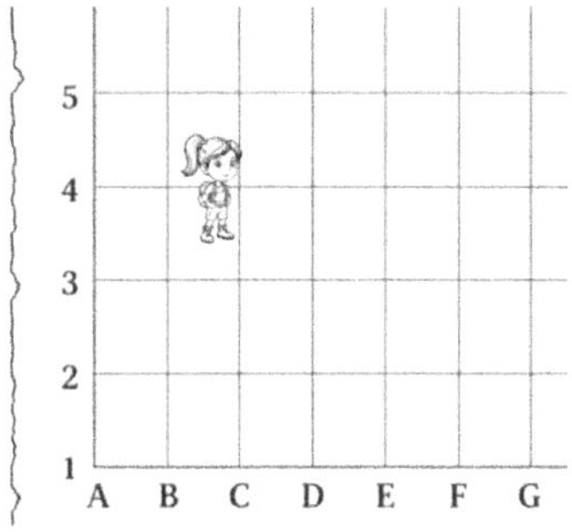

Her moves are:

north 14 → east 13 → south 12 → west 11 → north 10 → east 9 → south 8 → west 7 → north 6 → east 5 → south 4 → west 3 → north 2 → east 1

Which square does Mika finish on?

LET'S UNDERSTAND THE QUESTION TOGETHER

Mika begins at **C4** on the treasure map.

She moves in a repeating order:
north, then east, then south, then west.

Each instruction tells her how many squares to move in that direction.

We only need to find her **final square** after all the moves are finished.

A good way to think about this is:

- ✓ **North and south** change the number.
- ✓ **East and west** change the letter.

WHAT WE KNOW

Mika starts on **C4**

The directions repeat in this order:

- north
- east
- south
- west

Her moves are:

north 14 → east 13 → south 12 → west 11 → north 10 → east 9 → south 8 → west 7 → north 6 → east 5 → south 4 → west 3 → north 2 → east 1

WHAT WE NEED TO FIND

We need to find the **square where Mika finishes.**

Our Plan

We will separate the moves into:

- ✓ Horizontal moves: **east and west**
- ✓ Vertical moves: **north and south**

Then we will find the overall movement and apply it to **C4**.

Step-by-Step Solution

Step 1: Add the east moves.

East moves are:

$$13 + 9 + 5 + 1 = 28$$

So, Mika moves **28 squares east** in total.

Step 2: Add the west moves.

West moves are:

$11 + 7 + 3 = 21$

So, Mika moves **21 squares west** in total.

Step 3: Find the net horizontal movement.

$28 - 21 = 7$

So overall, Mika moves **7 squares east**.

Step 4: Work out the new column.

Mika starts at **C**.

Move 7 squares east:

- ✓ C to D = 1
- ✓ D to E = 2
- ✓ E to F = 3
- ✓ F to G = 4
- ✓ G to H = 5
- ✓ H to I = 6
- ✓ I to J = 7

So, the final column is **J**.

Step 5: Add the north moves.

North moves are:

$14 + 10 + 6 + 2 = 32$

So, Mika moves **32 squares north** in total.

Step 6: Add the south moves.

South moves are:

$12 + 8 + 4 = 24$

So, Mika moves **24 squares south** in total.

Step 7: Find the net vertical movement.

$$32 - 24 = 8$$

So overall, Mika moves **8 squares north**.

Step 8: Work out the new row.

Mika starts at row **4**.

Move 8 squares north:

$$4 + 8 = 12$$

So, the final row is **12**.

Step 9: Combine the column and row.

Final column = **J**

Final row = **12**

So, Mika finishes on **J12**.

Quick Check

East is greater than west, so Mika should end up to the **right** of C.

North is greater than south, so Mika should end up **above** 4.

Moving **7 squares east** from C gives **J**.

Moving **8 squares north** from 4 gives **12**.

So **J12** makes sense.

Final Answer

Mika finishes on **J12**.

Common Mistake

A common mistake is mixing up which part of the grid changes.

Remember:

- ✓ ***East and west** change the **letter***
- ✓ ***North and south** change the **number***

Exam Tip

In long movement questions, do not track every move one by one unless you need to. It is usually faster to total the horizontal moves and vertical moves separately first.

Time, Money & Rates

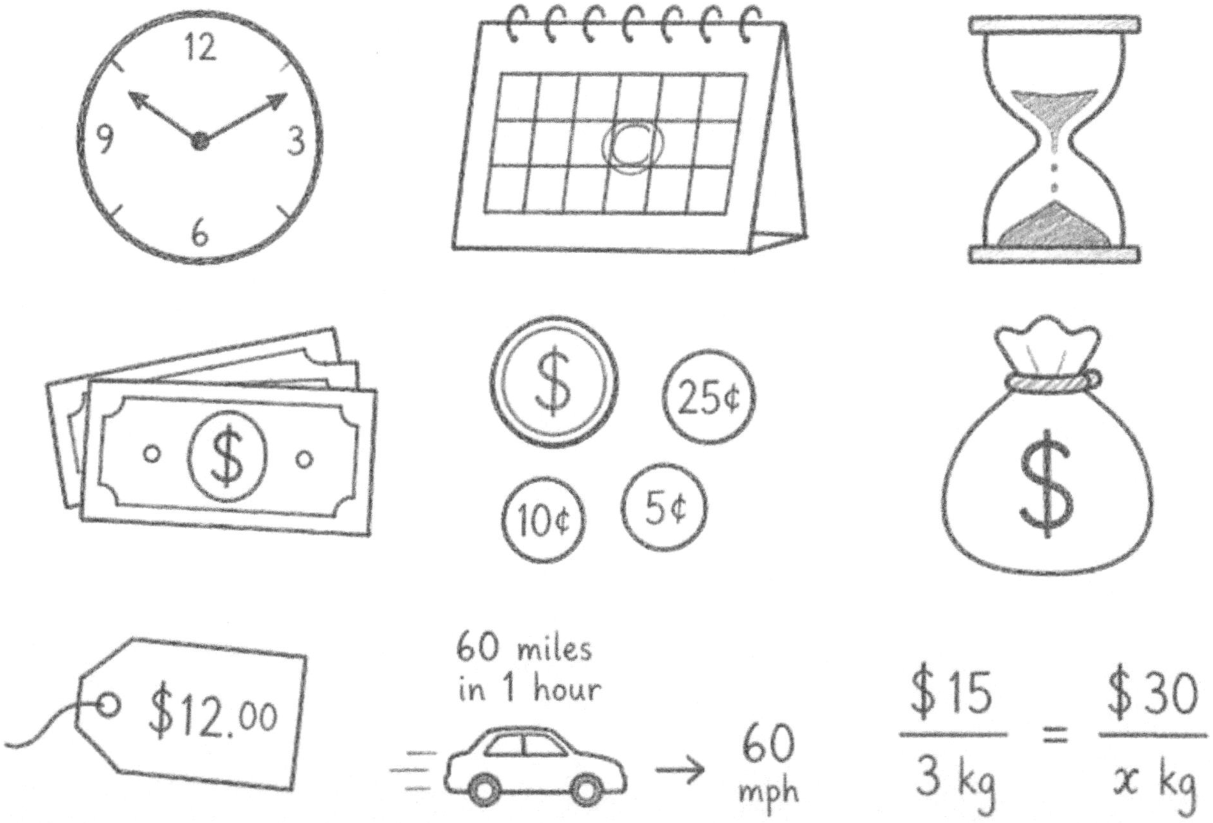

Elapsed Time with Fractions

QUESTION

Mika was baking cakes in her kitchen. She spent **$6\frac{3}{4}$** hours baking and cleaning up. She finished at **5:10 pm**.

What time did she start baking?

LET'S UNDERSTAND THE QUESTION TOGETHER

Mika spent a total of **$6\frac{3}{4}$ hours** baking.

She **finished at 5:10 pm**.

We need to find the time she **started**.

So, we are working **backwards** from the finishing time to the starting time.

WHAT WE KNOW

Mika spent **$6\frac{3}{4}$ hours** baking and cleaning up.

She finished at **5:10 pm**.

We need to find her **starting time**.

WHAT WE NEED TO FIND

We need to find what time Mika started baking.

Our Plan

First, change $\mathbf{6\frac{3}{4}}$ **hours** into **hours and minutes.**

Then subtract that amount of time from **5:10 pm**.

Good to Know

A simple way to remember common fractions of an hour:

- ✓ $\frac{1}{4}$ hour = **15 minutes**
- ✓ $\frac{1}{2}$ hour = **30 minutes**
- ✓ $\frac{3}{4}$ hour = **45 minutes**
- ✓ **1** hour = **60 minutes**

This works because **1 hour = 60 minutes.**

So:

- ✓ $\frac{1}{4}$ of 60 = 15
- ✓ $\frac{1}{2}$ of 60 = 30
- ✓ $\frac{3}{4}$ of 60 = 45

Step-by-Step Solution

Step 1: First, use the simple fraction-of-an-hour pattern.

We know: $\frac{3}{4}$ hour = **45 minutes**

So straight away: $\mathbf{6\frac{3}{4}}$ hours = 6 hours 45 minutes

If you want to do it **manually**, you can also work out $\frac{3}{4}$ of 60 minutes:

$$\frac{3}{4} \times 60 = 45$$

So: $6\frac{3}{4}$ hours = 6 hours 45 minutes.

Step 2: Write the finishing time.

Mika finished at **5:10 pm**.

Step 3: Subtract the **45 minutes** first.

$5{:}10 \text{ pm} - 45 \text{ minutes} = 4{:}25 \text{ pm}$

Step 4: Now subtract the **6 hours**.

$4{:}25 \text{ pm} - 6 \text{ hours} = 10{:}25 \text{ am}$

Step 5: State what the answer means.

Mika started baking at **10:25 am**.

Quick Check

From **10:25 am to 4:25 pm** is **6 hours**.

From **4:25 pm to 5:10 pm** is **45 minutes**.

So, the total time is **6 hours 45 minutes**, which matches **$6\frac{3}{4}$ hours**.

Final Answer

Mika started baking at **10:25 am**.

Common Mistake

*Don't add the time. Since Mika **finished** at 5:10 pm, we must **subtract** to work backwards.*

Exam Tip

*When you see a time question with words like **finished**, **ended**, or **left**, it usually means you need to **subtract** to find the starting time.*

Time zone calculations

QUESTION

Nai is video calling a cousin in Vancouver. Auckland's time zone is 20 hours ahead of Vancouver's time zone. What is the time and date in Vancouver?

LET'S UNDERSTAND THE QUESTION TOGETHER

We are comparing the time in **two different cities**.

Auckland is **20 hours ahead**, so Vancouver is **behind** Auckland.

That means when it is evening in Auckland, it could be much earlier in Vancouver.

We need to work out **both**:

- ✓ The **time**.
- ✓ The **date**.

So, we must be careful to see whether going back changes the **day** as well.

WHAT WE KNOW

- ✓ Auckland is **20 hours ahead** of Vancouver.
- ✓ The time in Auckland is **7:15 pm**.
- ✓ The date in Auckland is **3 March.**

WHAT WE NEED TO FIND

The **time and date in Vancouver**.

Good to Know: How to Convert from pm to 24-hour Time

For **pm** times, add **12** to the hour number, unless it is **12 pm**.

Examples:

- ✓ 1:00 pm = 13:00
- ✓ 4:30 pm = 16:30
- ✓ 7:15 pm = 19:15

So, in this question:

7:15 pm = 19:15

Our Plan

Since Auckland is **ahead**, we need to **go backwards 20 hours** from the Auckland time.

Then we check whether the date changes.

Step-by-Step Solution

Step 1: Write the Auckland time clearly.

Auckland time = **7:15 pm on 3 March**

Step 2: Convert the time to 24-hour time.

7:15 pm = **19:15**

Step 3: Go back 20 hours.

19:15 − 20 hours = **23:15** on the **previous day**

Step 4: Change 23:15 back to 12-hour time.

23:15 = **11:15 pm**

Step 5: Work out the date.

Because we went back past midnight, the date becomes **the day before 3 March**.

So, the date is **2 March**.

Quick Check

Auckland is much farther ahead than Vancouver, so Vancouver should be on an earlier time and possibly an earlier date.

Going back **20 hours** from **7:15 pm** gives **11:15 pm the previous day**, which fits.

Final Answer

The time in Vancouver is **11:15 pm on 2 March**.

Common Mistake

A common mistake is to subtract the hours correctly but forget that going past midnight changes the ***date.***

Exam Tip

When a question says ***ahead*** *or* ***behind,*** *decide first whether you need to* ***add*** *or* ***subtract.*** *Then check carefully whether the* ***day changes*** *too.*

Time Differences Between Cities

QUESTION

Mika is checking world times for a school event.

The time in Seoul is **2 hours ahead** of the time in Singapore.

The time in Nairobi is **5 hours behind** the time in Singapore.

When it is **4 pm in Nairobi**, what time is it in Seoul?

LET'S UNDERSTAND THE QUESTION TOGETHER

We are comparing the times in **three different places**.

We are told how **Seoul** and **Nairobi** each relate to the time in **Singapore**.

So, Singapore is the **middle reference time**.

The question gives us a time in **Nairobi** and asks for the matching time in **Seoul** at that same moment.

We need to carefully move from **Nairobi → Singapore → Seoul**.

WHAT WE KNOW

- ✓ Seoul is **2 hours ahead** of Singapore.
- ✓ Nairobi is **5 hours behind** Singapore.
- ✓ The time in Nairobi is **4 pm**.

WHAT WE NEED TO FIND

We need to find the time in **Seoul** when it is **4 pm in Nairobi**.

Our Plan

First, work out the time in **Singapore** using the Nairobi time.

Then use the Singapore time to find the time in **Seoul**.

Step-by-Step Solution

Step 1: Start with the time in Nairobi.

Nairobi time = **4 pm**

Step 2: Find the time in Singapore.

Nairobi is **5 hours behind** Singapore, so Singapore is **5 hours ahead** of Nairobi.

4 pm + 5 hours = 9 pm

So, the time in Singapore is **9 pm**.

Step 3: Find the time in Seoul.

Seoul is **2 hours ahead** of Singapore.

9 pm + 2 hours = 11 pm

So, the time in Seoul is **11 pm**.

Quick Check

Nairobi is 5 hours behind Singapore, and Seoul is 2 hours ahead of Singapore.
That means Seoul is **7 hours ahead of Nairobi** altogether.

4 pm + 7 hours = 11 pm

That matches our answer.

Final Answer

When it is **4 pm in Nairobi**, it is **11 pm in Seoul**.

Common Mistake

*A common mistake is mixing up **ahead** and **behind**.*
*If a place is **behind**, its time is earlier. If a place is **ahead**, its time is later.*

Exam Tip

*When a question has three places, use the city in the middle as your **reference point** and move one step at a time.*

Converting Times Across Time Zones

QUESTION

Mika is calling a cousin in London. London is **11 hours behind** Singapore time. A school concert starts in Singapore at **9:00 am on Monday**.

What time is it in London when the concert begins?

LET'S UNDERSTAND THE QUESTION TOGETHER

This question is about **time difference** between two places.

We are told:

- ✓ The concert starts in **Singapore**.
- ✓ London is **11 hours behind** Singapore.
- ✓ The starting time is **9:00 am on Monday.**

So, we need to work out what the **matching time in London** is at that exact moment.

Because London is **behind**, its time will be **earlier** than Singapore's time.

WHAT WE KNOW

Singapore time = **9:00 am Monday**.

London is **11 hours behind** Singapore.

We need the time in **London.**

WHAT WE NEED TO FIND

We need to find the **London time** when it is **9:00 am Monday in Singapore**.

Our Plan

We will subtract **11 hours** from the Singapore time, then check whether the **day changes**.

Step-by-Step Solution

Step 1: Write the starting time.

Singapore time is **9:00 am Monday**.

Step 2: Subtract the 11-hour time difference.

Since London is **11 hours behind**, we do:

9:00 am − 11 hours

Step 3: Subtract 9 hours first to reach midnight.

9:00 am Monday − 9 hours = **12:00 am Monday**

Step 4: Subtract the remaining 2 hours.

12:00 am Monday − 2 hours = **10:00 pm Sunday**

Step 5: State what this means.

When the concert starts in Singapore, the time in London is **10:00 pm on Sunday**.

Quick Check

Singapore is **ahead**, so London should have an **earlier time**.
Starting from **9:00 am Monday** and going back **11 hours** gives **10:00 pm Sunday**, which makes sense.

Final Answer

The time in London will be **10:00 pm on Sunday**.

Common Mistake

*A common mistake is **adding** the time difference instead of subtracting it.*
*Because London is **behind**, we must go **backward** in time.*

Exam Tip

*When solving time-zone questions, always circle the words **ahead** or **behind** first. That tells you whether to **add** or **subtract**.*

Converting Months to Seconds

QUESTION

Nai is recording how long a display screen stays on at a museum.

The screen will stay on for **15 months**.

Assume **1 month = 30 days**.

How many seconds is 15 months?

LET'S UNDERSTAND THE QUESTION TOGETHER

This question is about changing **months** into **seconds**.

We are starting with **15 months**.

But months are not directly connected to seconds, so we need to move through smaller time units step by step.

We are told to use **1 month = 30 days**. That is important because real months have different numbers of days, but this question gives us one fixed value to use.

WHAT WE KNOW

- ✓ 1 month = 30 days.
- ✓ 1 day = 24 hours.
- ✓ 1 hour = 60 minutes.
- ✓ 1 minute = 60 seconds.
- ✓ Total time = **15 months.**

WHAT WE NEED TO FIND

A **calculation** that shows how to convert **15 months into seconds**.

Our Plan

We will change:

- ✓ Months to days.
- ✓ Days to hours.
- ✓ Hours to minutes.
- ✓ Minutes to seconds.

Then we will write it as one multiplication calculation.

Step-by-Step Solution

Step 1: Convert months to days

15 months = 15×30 days

Step 2: Convert days to hours

Multiply by 24.

So now we have:

$15 \times 30 \times 24$

Step 3: Convert hours to minutes

Multiply by 60.

Now we have:
$15 \times 30 \times 24 \times 60$

Step 4: Convert minutes to seconds

Multiply by 60 again.

So, the full calculation is:

$\mathbf{15 \times 30 \times 24 \times 60 \times 60}$

Step 5: Work out the total

$15 \times 30 = 450$

$450 \times 24 = 10{,}800$

$10{,}800 \times 60 = 648{,}000$

$648{,}000 \times 60 = 38{,}880{,}000$

Quick Check

15 months is about 450 days.

Each day has 86,400 seconds.

$450 \times 86{,}400 = 38{,}880{,}000$, so our answer matches.

Final Answer

38,880,000 seconds.

Common Mistake

Using ***4 weeks in a month*** *instead of the given* ***30 days in a month****. Always use the information written in the question.*

Exam Tip

For big time conversions, write the units in order first:
months → days → hours → minutes → seconds
That helps you see exactly what to multiply by.

Comparing Unit Price and Pack Cost

QUESTION

It is **back-to-school time**, and Nai is buying supplies.

A pack of 10 pencils costs **$8.50**. Single pencils cost **$0.75 each**.

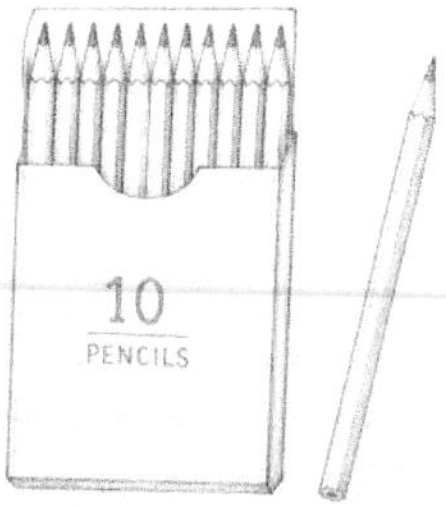

Nai buys **3 packs of pencils**.
How much more does Nai pay by buying packs instead of buying the same number of pencils individually?

LET'S UNDERSTAND THE QUESTION TOGETHER

Nai is getting ready for school and needs pencils.

There are **two ways** to buy them:

- ✓ In **packs of 10**.
- ✓ or as **single pencils**.

Nai buys **3 packs**, so we first need to know how many pencils that is altogether.

Then we compare this with buying the **same number of pencils individually**.

We are finding **how much extra** Nai pays by choosing packs instead of single pencils.

WHAT WE KNOW

- ✓ 1 pack = **10 pencils**.
- ✓ 1 pack costs **$8.50**
- ✓ Nai buys **3 packs**.
- ✓ Single pencil costs **$0.75**

WHAT WE NEED TO FIND

The **difference in cost** between:

- ✓ Buying packs.
- ✓ Buying the same number of pencils individually.

Our Plan

1. Work out how many pencils Nai buys in total.
2. Find the total cost of the packs.
3. Find the cost of the same number of single pencils.
4. Subtract to find how much more the packs cost.

Step-by-Step Solution

Step 1: Find total number of pencils.

$3 \times 10 =$ **30 pencils**

Step 2: Find total cost of packs.

$3 \times \$8.50 =$ **$25.50**

Step 3: Find cost of buying 30 single pencils.

$30 \times \$0.75 =$ **$22.50**

Step 4: Find how much more the packs cost.

$\$25.50 - \$22.50 =$ **$3.00**

Quick Check

Each pack pencil costs:

$\$8.50 \div 10 =$ **$0.85 per pencil**

Single pencil = **$0.75**

Difference per pencil = $0.10

For 30 pencils: $30 \times \$0.10 =$ **$3.00**

Final Answer

Nai pays **$3.00 more** by buying the packs.

Common Mistake

Forgetting to match the ***total number of pencils*** *before comparing costs. Always compare equal amounts.*

Exam Tip

Before you subtract, check that you are comparing ***the same total number of items****. If the amounts are not equal, your answer will be wrong even if your calculations are correct.*

Cost Savings (Bulk Purchase)

QUESTION

Nai needs to buy 5 kg of flower petals.

How much money will she save by buying 1 kg bags rather than 250 g bags?

LET'S UNDERSTAND THE QUESTION TOGETHER

Nai needs a total of **5 kg** of flower petals.

The petals are sold in **two different bag sizes**.

We need to compare:

- ✓ The total cost using only **250 g bags**.
- ✓ The total cost using only **1 kg bags**.

Then we find the **difference in cost**, which tells us the saving.

WHAT WE KNOW

- ✓ Nai needs **5 kg** of flower petals.
- ✓ **250 g bag = $1.25**
- ✓ **1 kg bag = $4.40**
- ✓ We are finding the **saving.**

WHAT WE NEED TO FIND

How much cheaper it is to buy **1 kg bags** instead of **250 g bags**.

Our Plan

1. Convert everything to the same unit.
2. Find how many bags are needed for each option.
3. Work out both total costs.
4. Subtract to find the saving.

Step-by-Step Solution

Step 1: Convert 5 kg into grams.

$5 \text{ kg} = 5000 \text{ g}$

Step 2: Find how many 250 g bags are needed.

$5000 \div 250 = 20$

So, Nai needs **20 small bags**.

Step 3: Find the total cost of small bags.

$20 \times \$1.25 = \$25.00$

Step 4: Find the total cost of the 1 kg bags.

$5 \times \$4.40 = \$22.00$

Step 5: Find the saving.

$\$25.00 - \$22.00 = \$3.00$

Quick Check

- ✓ 20×250 g $= 5000$ g
- ✓ 5×1 kg $= 5$ kg
- ✓ $\$25 - \$22 = \$3$

Everything matches!

Final Answer

Nai will save **$3.00** by buying the **1 kg bags**.

Alternative Method: Unit Price

First, convert 250 g to kilograms:
250 g = 0.25 kg

Now find the cost per kilogram for each option:

- **250 g bags:**
  $\$1.25 \div 0.25 = \$5.00$ per kg
- **1 kg bags:**
  $4.40 per kg

Now compare the prices:

- **Saving per kg:**
  $\$5.00 - \$4.40 = \$0.60$

For 5 kg:

- **Total saving:**
  $\$0.60 \times 5 = \$3.00$

So, Nai saves $3.00.

📌 **Common Mistake**

*Forgetting to convert **kg to grams** before working with smaller bag sizes can lead to the wrong number of bags.*

💡 **Exam Tip**

*When comparing deals, always calculate the **total cost for the exact amount needed** — not just the price per bag!*

Speed Calculation (Travel Rate)

QUESTION

Nai is rowing across a lake to a campsite. She wants to get there in 20 minutes.

What speed, in kilometres per hour, must Nai row at to arrive on time?

LET'S UNDERSTAND THE QUESTION TOGETHER

Nai needs to travel **12 km** across the lake.

She has only **20 minutes** to complete the trip.

The question is asking for her **speed**.

We also need to pay attention to the units. The distance is in **kilometres**, but the time is in **minutes**. The answer must be in **kilometres per hour**.

So, we are finding how fast Nai must row to cover the full distance in the given time.

WHAT WE KNOW

- ✓ Distance = **12 km**.
- ✓ Time = **20 minutes.**
- ✓ Speed must be in **km/h.**

WHAT WE NEED TO FIND

We need to find Nai's rowing speed in **kilometres per hour**.

Our Plan

1. We will use the rule: **speed = distance ÷ time**
2. First, we will convert **20 minutes** into **hours**.
3. Then we will divide the distance by the time.

Step-by-Step Solution

Step 1: Change 20 minutes into hours.

There are **60 minutes** in 1 hour.

$$20 \text{ minutes} = \frac{20}{60} \text{ hour} = \frac{1}{3} \text{ hour}$$

So, the rowing time is $\frac{1}{3}$ **hour.**

Step 2: Write the speed formula.

$$\text{speed} = \frac{\text{distance}}{\text{time}}$$

Step 3: Substitute the values.

$$\text{speed} = \frac{12}{\frac{1}{3}}$$

Step 4: Divide by $\frac{1}{3}$.

Dividing by $\frac{1}{3}$ is the same as multiplying by 3.

$$12 \div \frac{1}{3} = 12 \times 3 = 36$$

Step 5: Write the unit.

36 km/h

Quick Check

At **36 km/h**, in **1 hour** Nai would row **36 km**.

In $\frac{1}{3}$ **of an hour**:

$$36 \div 3 = 12$$

That matches the distance, so the answer is correct.

Final Answer

Nai must row at **36 kilometres per hour** to arrive on time.

Alternative Method

20 minutes is one-third of an hour.

If Nai rows 12 km in one-third of an hour, then in 1 hour she would row:

$$12 \times 3 = 36\text{km}$$

So, her speed is **36 km/h**.

Common Mistake

*A common mistake is forgetting to change **20 minutes** into **hours** before dividing.*

Exam Tip

Using $\frac{1}{3}$ *is usually the best exam method because it is exact and keeps the working neat.*
*You could also use the decimal way: **20 minutes = 0.333... hours**, but that decimal goes on forever, so it is easier to make a mistake. Fractions are often safer.*

Systems of Equations (Token Values)

QUESTION

Mika collects two types of tokens at an arcade. Each **black token** is worth 3 points and each **grey token** is worth 6 points. Mika has the **same number of black and grey tokens**, and the total number of points is **162**.

Each black token weighs **4 grams**, and each grey token weighs **10 grams**.

What is the **total weight** of all the tokens Mika has?

LET'S UNDERSTAND THE QUESTION TOGETHER

Mika has **two types of tokens**: black and grey.

- ✓ There are **equal numbers** of each type.
- ✓ Each type has a different **value** and **weight**.
- ✓ The total value of all tokens is **162 points**.

We need to find the **combined weight** of all the tokens.

WHAT WE KNOW

- ✓ Number of black tokens = number of grey tokens.
- ✓ Black token value = 3 points.
- ✓ Grey token value = 6 points.
- ✓ Total value = 162 points.
- ✓ Black token weight = 4 g.
- ✓ Grey token weight = 10 g.

WHAT WE NEED TO FIND

The **total weight** of all the tokens.

Our Plan

1. Let the number of each type be the same (call it n).
2. Use the total value to find n.
3. Work out the weight of each type.
4. Add them together.

Step-by-Step Solution

Step 1: Since they are equal, each type has $\boldsymbol{n}$ tokens.

Step 2: Write the total value equation

$$3n + 6n = 162$$

Step 3: Simplify and solve

$$9n = 162$$
$$n = 162 \div 9 = 18$$

So, there are **18 black tokens and 18 grey tokens**.

Step 4: Find total weight of black tokens

$$18 \times 4 = 72 \text{ g}$$

Step 5: Find total weight of grey tokens

$$18 \times 10 = 180 \text{ g}$$

Step 6: Add the weights

$$72 + 180 = 252 \text{ g}$$

Quick Check

- ✓ Black value: $18 \times 3 = 54$
- ✓ Grey value: $18 \times 6 = 108$
- ✓ Total: $54 + 108 = 162$

Final Answer

The total weight of all the tokens is **252 grams**.

Alternative Method

Since Mika has the same number of black and grey tokens, we can group them into pairs.

Each pair contains:

- ✓ 1 black token and 1 grey token

Step 1: Find the value of one pair

$3 + 6 = 9\text{points}$

Step 2: Find the number of pairs

$162 \div 9 = 18\text{pairs}$

Step 3: Find the weight of one pair

$4 + 10 = 14\text{g}$

Step 4: Find the total weight

$18 \times 14 = 252\text{g}$

So, the total weight is 252 grams.

Common Mistake

*Mixing up the values and weights, or forgetting that the numbers of tokens are **equal**.*

Exam Tip

*When you see **"same number of each"**, use one variable (like n) for both. It makes forming equations much easier!*

Counting Coin Combinations (Change)

QUESTION

At the school canteen, Mika buys lunch for **$4.30** and pays with a **$6 note**.

The cashier can give change using only **$1 coins, 50 cent coins, and 10 cent coins**.

In how many ways can the cashier combine these coins to give Mika the correct change?

LET'S UNDERSTAND THE QUESTION TOGETHER

Mika pays **more** than the lunch costs, so she will get **change** back.

First, we need to find **how much change** she should receive.

Then we need to count all the different **coin combinations** that make exactly that amount.

The cashier can only use these coins:

- ✓ **$1 coins**.
- ✓ **50 cent coins**.
- ✓ **10 cent coins**.

We are counting different **combinations**, so the **order** of the coins does not matter.

WHAT WE KNOW

- ✓ Mika pays **$6.00**
- ✓ The lunch costs **$4.30**
- ✓ The change must be given using only:
    - ✓ **$1 coins**.
    - ✓ **50 cent coins**.
    - ✓ **10 cent coins**.

WHAT WE NEED TO FIND

We need to find **how many different combinations of these coins** make the correct change.

Our Plan

1. First, work out the change.
2. Then write an equation in **cents** using the coin values.
3. After that, start with the **largest coin** and test all possible cases in an organised way.
4. Finally, count the combinations that work.

Step-by-Step Solution

Step 1: Find the change.

$\$6.00 - \$4.30 = \$1.70$

So, the cashier must give **$1.70**, which is **170 cents.**

Step 2: Write the value of each coin in cents.

The coins allowed are:

- ✓ **$1 coin = 100 cents.**
- ✓ **50 cent coin = 50 cents.**
- ✓ **10 cent coin = 10 cents.**

We will use letters to represent the number of each coin:

- ✓ $a=$ number of **$1 coins.**
- ✓ $b=$ number of **50 cent coins.**
- ✓ $c=$ number of **10 cent coins.**

So, the total value of the coins is:

$$100a + 50b + 10c$$

Since the coins must add up to **170 cents**, we get:

$$100a + 50b + 10c = 170$$

This equation represents all the possible coin combinations.

Step 3: Choose the possible numbers of $1 coins.

A $1 coin is worth **100 cents**.

The total change is **170 cents**, so the cashier can use:

- ✓ **0** one-dollar coins, or
- ✓ **1** one-dollar coin.

The cashier cannot use **2** one-dollar coins because that would be **200 cents**, which is too much.

Step 4: Case 1: Use 0 one-dollar coins.

Substitute $a = 0$ into the equation:

$$100(0) + 50b + 10c = 170$$
$$50b + 10c = 170$$

Divide by 10 to make it simpler:

$$5b + c = 17$$

Now test possible values of b:

- ✓ If $b = 0$, then $c = 17$
- ✓ If $b = 1$, then $c = 12$
- ✓ If $b = 2$, then $c = 7$
- ✓ If $b = 3$, then $c = 2$

If $b = 4$, then:

$$5b = 20$$

This exceeds 17, so no further values are possible.

So, this case gives **4 combinations.**

Step 5: Case 2: Use 1 one-dollar coin.

Substitute $a = 1$ into the equation:

$$100(1) + 50b + 10c = 170$$
$$100 + 50b + 10c = 170$$

Subtract 100 from both sides:

$$50b + 10c = 70$$

Divide by 10:

$$5b + c = 7$$

Now test possible values of b:

- ✓ If $b = 0$, then $c = 7$
- ✓ If $b = 1$, then $c = 2$

If $b = 2$, then:

$$5b = 10$$

That is too much.

So, this case gives **2 combinations.**

Step 6: Add the combinations from both cases.

$$4 + 2 = 6$$

So, there are **6 different ways** to give the correct change.

Step 7: List the 6 combinations.

1. **17** ten-cent coins.
2. **1** fifty-cent coin and **12** ten-cent coins.
3. **2** fifty-cent coins and **7** ten-cent coins.
4. **3** fifty-cent coins and **2** ten-cent coins.
5. **1** one-dollar coin and **7** ten-cent coins.
6. **1** one-dollar coin, **1** fifty-cent coin, and **2** ten-cent coins.

Quick Check

Let's check one combination:

$\$1.00 + \$0.50 + \$0.20 = \$1.70$

That matches the change exactly.

We also checked all possible numbers of **$1 coins**, so we did not miss any combinations.

Final Answer

The cashier can combine the coins in **6 different ways** to give Mika the correct change.

📌 Common Mistake

A common mistake is to count the same combination more than once in a different order.

For example,
$1, 50c, 10c, 10c *is the same combination as*
10c, $1, 10c, 50c*.*

The order does ***not*** *matter.*

💡 Exam Tip

In coin combination questions, always start with the ***largest coin*** *first.*
This helps you work in a neat, organised way and makes it easier to count every valid combination.

Arithmetic Sequence (reading time)

QUESTION

Mika is reading 4 chapters of a book.

The first chapter takes **18 minutes** to read.

Each next chapter takes **2 minutes less** than the chapter before it.

How many minutes does Mika spend reading all 4 chapters?

LET'S UNDERSTAND THE QUESTION TOGETHER

We are told about **4 chapters**.

The **first chapter** takes **18 minutes**.

After that, each chapter is **2 minutes shorter** than the one before.

We need to find the **total reading time** for **all 4 chapters together**.

WHAT WE KNOW

- ✓ There are **4 chapters**.
- ✓ The first chapter takes **18 minutes**.
- ✓ Each next chapter takes **2 minutes less** than the previous one.

WHAT WE NEED TO FIND

We need to find the **total number of minutes** for all 4 chapters.

Our Plan

First, work out how long each chapter takes.

Then add the 4 times together.

Step-by-Step Solution

Step 1: Write the time for the first chapter.

First chapter = **18 minutes**

Step 2: Work out the second chapter.

It is 2 minutes less than 18.
$18 - 2 = 16$
Second chapter = **16 minutes**

Step 3: Work out the third chapter.

It is 2 minutes less than 16.
$16 - 2 = 14$
Third chapter = **14 minutes**

Step 4: Work out the fourth chapter.

It is 2 minutes less than 14.
$14 - 2 = 12$
Fourth chapter = **12 minutes**

Step 5: Add all the chapter times.

$18 + 16 + 14 + 12 = 60$

Total reading time = **60 minutes**

Quick Check

The times are 18, 16, 14, and 12.

Adding them again:
$18 + 16 = 34$
$14 + 12 = 26$
$34 + 26 = 60$

So, the total is correct.

Final Answer

Mika spends **60 minutes** reading all 4 chapters.

Common Mistake

A common mistake is only subtracting once and forgetting to keep finding the time for ***each*** *chapter before adding them all.*

Exam Tip

When a question says each new item is ***shorter*** *or* ***less*** *than the one before, list the values one by one first. This helps you stay organised and avoid missing one.*

Rates and Resource Consumption (fish feeding)

QUESTION

Mika has a container of fish food that is completely full.

The food in the container is enough to feed:

- ✓ **1 large fish every day for 4 days**, or
- ✓ **1 small fish every day for 8 days**

What is the greatest number of days the container can feed **1 large fish and 1 small fish together**?

LET'S UNDERSTAND THE QUESTION TOGETHER

We are told the container starts **full**.

The same full container can be used in **two different ways**:

- ✓ For **1 large fish** for **4 days**.
- ✓ For **1 small fish** for **8 days.**

This means the **large fish eats more each day** than the small fish.

We need to find how long the food will last when **both fish are fed every day**.

So, the question is asking for the **maximum whole number of days** the full container can last for **both fishe together**.

WHAT WE KNOW

- ✓ A full container feeds **1 large fish for 4 days**.
- ✓ A full container feeds **1 small fish for 8 days**.
- ✓ Both fish are fed **every day**.
- ✓ We need the **largest number of days** the container will last.

WHAT WE NEED TO FIND

We need to find the greatest number of days that **1 large fish and 1 small fish together** can be fed from the full container.

Our Plan

1. First, compare how much food each fish eats in one day.
2. Then work out how much food they eat **together in one day**.
3. Finally, use the full container amount to find how many days it lasts.

Step-by-Step Solution

Step 1: Use the full container as one whole amount.

Let the full container be **8 small-fish daily portions.**

We can do this because the container feeds **1 small fish for 8 days.**

So, the full container = **8 small-fish portions.**

Step 2: Work out how much 1 large fish eats each day.

The same full container also feeds **1 large fish for 4 days.**

So, each day, the large fish eats:

$$8 \div 4 = 2$$

So, **1 large fish eats 2 small-fish portions per day.**

Step 3: Work out how much both fish eat together each day.

✓ 1 large fish eats **2 portions**
✓ 1 small fish eats **1 portion**

Together, they eat:

$$2 + 1 = 3$$

So, both fish together eat **3 small-fish portions per day.**

Step 4: Find how many full days the container lasts.

The full container has **8 portions**.

Each day both fish use **3 portions**.

$$8 \div 3 = 2 \text{ remainder } 2$$

So, the food lasts for **2 full days**.

Quick Check

After 2 days, both fish would use:

$$2 \times 3 = 6 \text{ portions}$$

That fits inside the full container of 8 portions.

A 3rd full day would need:

$$3 \times 3 = 9 \text{ portions}$$

But 9 is more than 8, so **3 days is too many**.

Final Answer

The container can feed **1 large fish and 1 small fish together for 2 days.**

Common Mistake

A common mistake is to add the days first:

$$4 + 8 = 12$$

That does ***not*** *make sense here, because the two animals eat at* ***different daily rates****. We must compare how much each one eats* ***per day****.*

Exam Tip

When one full amount can be used in two different ways, turn it into ***daily amounts first****. That makes it much easier to combine the rates correctly.*

Multi-Step Fraction Problem (apples)

QUESTION

There were **90 apples** in a basket. Some apples were taken out to make fruit packs.

Then **half of the apples left** were given to Nai. Nai ate **one-fifth** of the apples **she** received. Nai ate **8 apples**.

How many apples were taken out to make the fruit packs?

LET'S UNDERSTAND THE QUESTION TOGETHER

We start with **90 apples** in the basket.

Then some apples are removed.

After that, **half of the remaining apples** are given to Nai.

We are told what Nai does next.

She eats **one-fifth** of the apples she got, and that amount is **8 apples**.

We need to find the number of apples that were **taken out earlier** to make the fruit packs.

WHAT WE KNOW

- ✓ There were **90 apples** at the start.
- ✓ Some apples were taken out to make fruit packs.
- ✓ Then **half of the apples left** were given to Nai.
- ✓ Nai ate **one-fifth** of the apples **she** received.
- ✓ Nai ate **8 apples.**

WHAT WE NEED TO FIND

We need to find **how many apples were taken out to make the fruit packs**.

Our Plan

1. We will work backwards.
2. First, find how many apples Nai received.
3. Then find how many apples were left in the basket before Nai got half.
4. Finally, compare that number with the original **90 apples**.

Step-by-Step Solution

Step 1: Find how many apples Nai received.

Nai ate **one-fifth** of the apples she received, and that was **8 apples**.

$8 \times 5 = 40$

So, Nai received **40 apples**.

Step 2: Find how many apples were left in the basket before Nai was given half.

Nai received **half** of the apples left.

If half was **40 apples**, then the whole amount left was:

$40 \times 2 = 80$

So, there were **80 apples left** after the fruit packs were made.

Step 3: Find how many apples were taken out earlier.

There were **90 apples** at the start and **80 apples** were left.

$90 - 80 = 10$

So, **10 apples** were taken out to make the fruit packs.

Quick Check

Let's test the answer.

- ✓ Start with **90 apples**
- ✓ Take out **10 apples** → $90 - 10 = 80$
- ✓ Give Nai half of 80 → $80 \div 2 = 40$
- ✓ Nai eats one-fifth of 40 → $40 \div 5 = 8$

Everything matches.

Final Answer

10 apples were taken out to make the fruit packs.

Common Mistake

*A common mistake is to think the **8 apples** Nai ate is the same as the number taken out at the start.*

*But the **8 apples** are only **one-fifth of what Nai received**, so we must work backwards first.*

Exam Tip

*When you see fractions like **half**, **one-fifth**, or **one-quarter**, stop and ask, "What whole amount does this smaller part come from?" That helps you work backwards carefully.*

Relative Time Rate (faulty clock)

QUESTION

Mika's tracker was showing the correct time until **9:20 am**. At that moment, it malfunctioned and began running **three times as fast as normal**.

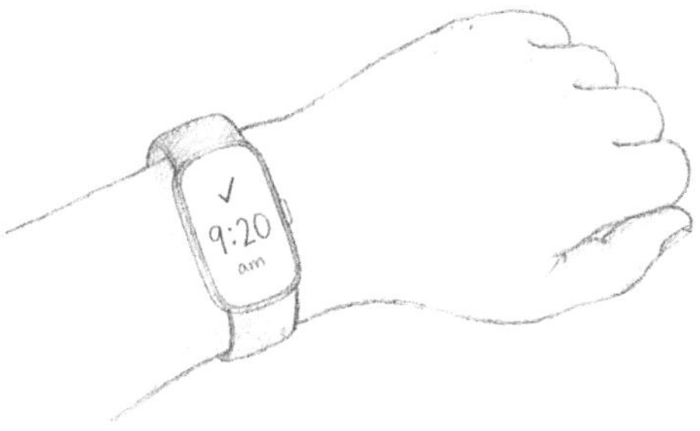

At **1:20 pm**, what time did the tracker show?

LET'S UNDERSTAND THE QUESTION TOGETHER

The tracker showed the correct time up to 9:20 am.

After that, it started moving faster than normal and now runs **3 times as fast**.

This means for every 1 hour in real life, the tracker moves 3 hours forward.

We are asked to find what time the tracker shows at 1:20 pm.

WHAT WE KNOW

- ✓ Correct time until **9:20 am.**
- ✓ Then it runs **3 times as fast.**
- ✓ Real time later is **1:20 pm.**

WHAT WE NEED TO FIND

The **time shown on the tracker** at **1:20 pm**.

Our Plan

1. Find the **real time passed** from 9:20 am to 1:20 pm.
2. Multiply that time by **3.**
3. Add the result to **9:20 am.**

Step-by-Step Solution

Step 1: Find the real time passed

From **9:20 am to 1:20 pm** is: **4 hours**

Step 2: Multiply by 3 (since it runs three times as fast)

4 hours × 3 = **12 hours**

So, the tracker moves forward **12 hours**.

Step 3: Add 12 hours to 9:20 am

9:20 am + 12 hours = **9:20 pm**

Quick Check

Real time = 4 hours

Triple speed → 12 hours. So, 9:20 am + 12 hours = 9:20 pm

Final Answer

At **1:20 pm**, the tracker shows **9:20 pm**.

Common Mistake

A common mistake is multiplying incorrectly or forgetting to multiply at all. Some students may multiply by the wrong number or forget to multiply at all.

Exam Tip

*Carefully read phrases like **"twice as fast"** or **"three times as fast"**. They tell you what number to multiply the real elapsed time by.*

Greatest Possible Difference (decimals)

QUESTION

Mika's little sister tore the corner of a page, hiding the last digit of two measurements. They now read **47.6? cm** and **42.9? cm**. Each? stands for one digit.

What is the **greatest possible difference** between the measurements?

LET'S UNDERSTAND THE QUESTION TOGETHER

There are **two measurements.**

In each measurement, the **last digit is missing**.

That means each? can be any digit from **0 to 9**.

We are asked for the **greatest possible difference**. That means we want the two measurements to be as **far apart** as possible.

WHAT WE KNOW

- ✓ One measurement is **47.6? cm**.
- ✓ The other measurement is **42.9? cm**.
- ✓ Each? stands for **one digit.**
- ✓ We need the **greatest possible difference.**

WHAT WE NEED TO FIND

We need to find the **largest possible difference** between the two measurements.

Our Plan

1. Make the **larger measurement as large as possible** and the **smaller measurement as small as possible**.
2. Then subtract.

Step-by-Step Solution

Step 1: Work out which measurement is already larger.

Compare **47.6?** and **42.9?**

Since 47 is greater than 42, **47.6?** cm is the larger measurement.

Step 2: Make the larger measurement as large as possible.

To make **47.6?** as large as possible, use **9** for the missing digit.

So, the measurement becomes: **47.69 cm**

Step 3: Make the smaller measurement as small as possible.

To make **42.9?** as small as possible, use **0** for the missing digit.

So, the measurement becomes: **42.90 cm**

Step 4: Subtract the smaller measurement from the larger measurement.

$47.69 - 42.90 = 4.79$

So, the greatest possible difference is **4.79 cm**.

Quick Check

Let's check if this makes sense.

- ✓ **47.69 cm** is the biggest possible first measurement.
- ✓ **42.90 cm** is the smallest possible second measurement.

So, this gives the **largest gap**.

Also, $47.7 - 42.9$ is about **4.8**, which matches **4.79 cm**.

Final Answer

The greatest possible difference between the measurements is **4.79 cm**.

Common Mistake

*A common mistake is to put **9** in both missing places.*

This does not give the greatest possible difference.

*For the **greatest difference**, make the **larger number bigger** and the **smaller number smaller**.*

Exam Tip

*To get the **greatest difference**, make the **bigger number as big as possible** by using **9**, and make the **smaller number as small as possible** by using **0**.*

Optimization of Bulk Purchasing (unit pricing)

QUESTION

Mika has **$26** to spend at the canteen. Using the prices shown in the image, what is **the greatest number** of fruit bars Mika can buy?

LET'S UNDERSTAND THE QUESTION TOGETHER

Mika wants to buy as **many fruit bars as possible** with **$26.**

There are **three buying choices**:

- ✓ Buying **single** fruit bars.
- ✓ Buying a **box of 6**
- ✓ Buying a **tray of 15**

The question is not asking for the cheapest way to buy one group.
It is asking for the **largest total number** of fruit bars Mika can get.

So, we need to think carefully about which combination gives the **most fruit bars** without spending more than **$26.**

WHAT WE KNOW

- ✓ 1 fruit bar = **$0.80**
- ✓ 6 fruit bars = **$4.20**
- ✓ 15 fruit bars = **$9.75**
- ✓ Mika has **$26.00**

WHAT WE NEED TO FIND

We need to find the **greatest number of fruit bars** Mika can buy for **$26 or less**.

Our Plan

1. First, compare which option gives the **best value per fruit bar**.
2. Then, start with the best-value option and check combinations to see which one gives the **largest total number**.

Step-by-Step Solution

Step 1: Work out the cost of 1 fruit bar in each option.

A single fruit bar costs **$0.80**.

A box of 6 costs **$4.20**, so each fruit bar costs:

$$\$4.20 \div 6 = \$0.70$$

A tray of 15 costs **$9.75**, so each fruit bar costs:

$$\$9.75 \div 15 = \$0.65$$

So, the **15-pack tray** is the cheapest way to buy each fruit bar.

Step 2: Buy as many 15-packs as possible first.

Each tray costs **$9.75**.

Two trays cost:

$$\$9.75 + \$9.75 = \$19.50$$

Mika has **$26.00**, so the money left is:

$$\$26.00 - \$19.50 = \$6.50$$

Step 3: Use the remaining **$6.50** as wisely as possible.

A box of 6 costs **$4.20**, so Mika can buy 1 box.

Money left after buying the box:

$6.50 - $4.20 = $2.30

A single fruit bar costs **$0.80**.

With **$2.30**, Mika can buy at **most 2 single** fruit bars because:

$2 \times \$0.80 = \$1.60$

but 3 single fruit bars would cost:

$3 \times \$0.80 = \$2.40$

and that is too much.

Step 4: Add the total number of fruit bars.

From 2 trays, Mika gets:

$15 + 15 =$ **30 fruit bars.**

From 1 box, Mika gets: **6 fruit bars.**

From the singles, Mika gets: **2 fruit bars.**

So, the **total number** of fruit bars is:

30 + 6 + 2 = 38 fruit bars

Step 5: Check the total cost.

The cost is:

- ✓ 2 trays = $19.50
- ✓ 1 box = $4.20
- ✓ 2 singles = $1.60

Total cost: $19.50 + $4.20 + $1.60 = $25.30

Since **$25.30 is less than $26.00**, this combination works.

Step 6: State the answer.

The greatest number of fruit bars Mika can buy is **38**.

Step 7: Check if another combination could give more fruit bars.

Try replacing one 15-pack with smaller packs.

A 15-pack gives 15 bars for \$9.75.

Using boxes instead:

$\$9.75 \div \$4.20 \approx 2$ boxes (12 bars), which is fewer than 15 bars.

Using only singles would give even fewer bars.

So, the 15-pack is still the best starting choice, and **no other combination gives more than 38 bars**.

Quick Check

Our final combination is:

2 trays + 1 box + 2 singles

That gives:

$15 + 15 + 6 + 2 = 38$ fruit bars

The total cost is:

$\$19.50 + \$4.20 + \$1.60 = \$25.30$

This is within Mika's **\$26 budget**, so the answer works.

Final Answer

Mika can buy **38 fruit bars**.

Common Mistake

A common mistake is to stop after buying the big packs and forget to use the leftover money on extra single fruit bars.

Exam Tip

*When a question asks for the **greatest number** you can buy, compare the **cost per item** first, then use any leftover money carefully to increase the total.*

Least Common Multiple (intervals)

QUESTION

If all three bells ring at the **same time now**, after how many minutes will they **ring together again**?

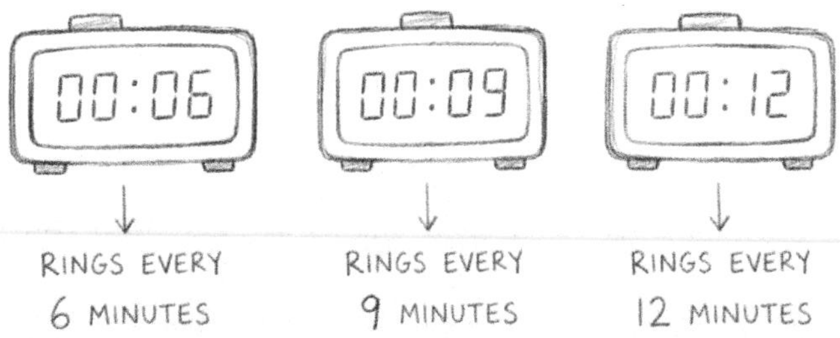

LET'S UNDERSTAND THE QUESTION TOGETHER

All three bells ring **at the same time right now**.

After that, each bell keeps ringing in its own pattern:

- ✓ One every **6 minutes**.
- ✓ One every **9 minutes**.
- ✓ One every **12 minutes.**

We need to find the **first time** when all three patterns match again.
So we are looking for the smallest number of minutes that is a multiple of all three intervals.

WHAT WE KNOW

- ✓ First bell: every **6 minutes**.
- ✓ Second bell: every **9 minutes.**
- ✓ Third bell: every **12 minutes**.
- ✓ They all ring together now.

WHAT WE NEED TO FIND

The **first number of minutes** after now when all three bells ring together again.

Our Plan

We will find the **LCM** of 6, 9, and 12.

LCM means **Lowest Common Multiple**. It is the **smallest number** that all three numbers divide into exactly (**with no remainder**).

We use **LCM** when we need to find the **first number or time that is common to all the patterns**. In this question, each clock chimes at its own interval, so we need the **first time all three intervals match**.

Step-by-Step Solution

Step 1: Prime factorise each number

Write each interval as a product of prime numbers.

- ✓ $6 = \mathbf{2 \times 3}$
- ✓ $9 = \mathbf{3 \times 3 = 3^2}$
- ✓ $12 = \mathbf{2 \times 2 \times 3 = 2^2 \times 3}$

Step 2: List the different primes

The prime numbers involved are 2 and 3.

Step 3: Take the highest power of each prime

Now choose the greatest number of each prime needed.

- ✓ For **2**, the highest power is $\mathbf{2^2}$ from 12
- ✓ For **3**, the highest power is $\mathbf{3^2}$ from 9

So, the LCM is: $\mathbf{2^2 \times 3^2}$

Step 4: Multiply to find the LCM

First work out the powers:

- ✓ $2^2 = \mathbf{4}$
- ✓ $3^2 = \mathbf{9}$

Now multiply: $\mathbf{4 \times 9 = 36}$

So, the LCM is **36**.

Step 5: Say what the answer means

This means all three bells will ring together again in **36 minutes**.

Quick Check

Let's test 36:

✓ $36 \div 6 = \mathbf{6}$

✓ $36 \div 9 = \mathbf{4}$

✓ $36 \div 12 = \mathbf{3}$

All answers are whole numbers, so **36** works.

Final Answer

All three bells will ring together again after **36 minutes**.

Common Mistake

*A common mistake is to stop at **18** because 18 works for 6 and 9.*
*But 18 does **not** work for 12, so it is not the correct answer.*

Exam Tip

*When a question asks when repeating events happen **together again**, that is a strong clue to use **LCM**.*
*Look for the **smallest time** that fits all the intervals.*

Elapsed Time from Timetables (ferry trip)

QUESTION

A ferry runs between Circular Quay and Manly according to the timetable shown.

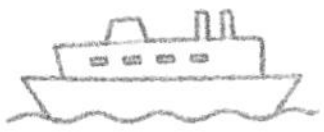

FERRY TIMETABLE

CIRCULAR QUAY to MANLY		MANLY to CIRCULAR QUAY	
DEPART	ARRIVE	DEPART	ARRIVE
8:10 am	9:00 am	7:30 am	8:20 am
11:25 am	12:15 pm	10:05 am	10:55 am
2:40 pm	3:30 pm	1:15 pm	2:05 pm
5:50 pm	6:40 pm	4:45 pm	5:35 pm

Mika catches the **third ferry** of the day from Manly to Circular Quay.
After arriving, Mika returns to Manly on the **next available ferry**.

How long is Mika **away** from Manly?

LET'S UNDERSTAND THE QUESTION TOGETHER

Mika starts in **Manly**.
Mika takes the **third ferry** going to **Circular Quay**.
After arriving there, Mika comes back on the **next ferry** heading to **Manly**.

We are finding the total time Mika is **away from Manly**.
That means we start timing when Mika **leaves Manly** and stop when Mika **arrives back in Manly**.

WHAT WE KNOW

- ✓ The **third ferry from Manly to Circular Quay** leaves at **1:15 pm**.
- ✓ It arrives at **Circular Quay at 2:05 pm**.
- ✓ The **next ferry from Circular Quay to Manly** leaves at **2:40 pm**.
- ✓ It arrives at **Manly at 3:30 pm.**

WHAT WE NEED TO FIND

We need to find the total amount of time Mika is away from Manly.

Our Plan

1. Find when Mika leaves Manly.
2. Find when Mika gets back to Manly.
3. Then work out the time between those two moments.

Step-by-Step Solution

Step 1: Find the ferry Mika takes from Manly.

The third ferry from **Manly to Circular Quay** departs at **1:15 pm**.

Step 2: Find when this ferry arrives.

This ferry arrives at **Circular Quay at 2:05 pm**.

This tells us when Mika is ready to catch the next ferry.

Step 3: Find the next ferry back to Manly.

After **2:05 pm**, the next ferry from **Circular Quay to Manly** departs at **2:40 pm** and arrives at **3:30 pm**.

Step 4: Work out how long Mika is away.

From **1:15 pm to 3:30 pm**:

- ✓ 1:15 pm → 2:15 pm = 1 hour
- ✓ 2:15 pm → 3:15 pm = 1 hour
- ✓ 3:15 pm → 3:30 pm = 15 minutes

Total = **2 hours 15 minutes**

Quick Check

Trip to Circular Quay: **50 minutes**

Waiting time: 2:05 pm to 2:40 pm = **35 minutes**

Trip back to Manly: **50 minutes**

Total time: 50 minutes + 35 minutes + 50 minutes = **135 minutes**

135 minutes = **2 hours 15 minutes.**

So, the answer is correct.

Final Answer

Mika is away from Manly for **2 hours 15 minutes**.

📌 Common Mistake

A common mistake is choosing the wrong return ferry. Be careful to use the ***next available ferry*** *after Mika arrives at Circular Quay.*

Exam Tip

In timetable questions, first mark the ***starting trip****, then the* ***next return trip****. After that, measure the full time from the first departure to the final arrival.*

Rates and Proportional Reasoning (rotations)

QUESTION

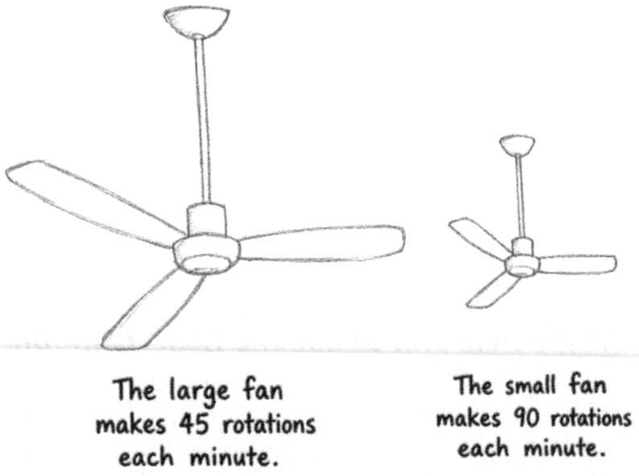

a) When the large fan has **turned 28 times**, how many times has **the small fan turned**?

b) If the large fan keeps turning for **80 minutes**, how many **rotations does it make altogether**?

LET'S UNDERSTAND THE QUESTION TOGETHER

This question is about **comparing rotations**.

The **small fan spins faster** than the large fan.

In part **a)**, both fans are turning for the **same amount of time**. We need to compare how many turns each fan makes in that time.

In part **b)**, we only look at the **large fan**. We need to find how many full rotations it makes after a certain number of minutes.

WHAT WE KNOW

- ✓ Large fan speed = **45 rotations per minute**.
- ✓ Small fan speed = **90 rotations per minute**.
- ✓ In part **a)**, the large fan turns **28 times**.
- ✓ In part **b)**, the large fan runs for **80 minutes**.

WHAT WE NEED TO FIND

We need to find:

- **a)** How many times the small fan turns.
- **b)** How many rotations the large fan makes in total.

Our Plan

We will compare the fan speeds first.

Then:

- ✓ For part **a)**, use the speed relationship between the two fans.
- ✓ For part **b)**, multiply the large fan's rotations per minute by the number of minutes.

Step-by-Step Solution

Part a)

Step 1: Compare the two fan speeds.

The small fan turns **90 rotations per minute** and the large fan turns **45 rotations per minute**.

$90 \div 45 = 2$

So, the small fan spins **twice as fast** as the large fan.

Step 2: Use this to find the small fan's rotations.

If the large fan **turns 28 times** in the same time, the small fan turns:

$28 \times 2 = 56$

Step 3: State the answer for part a.

The small fan turns **56 times**.

Part b)

Step 4: Work out the total number of rotations of the large fan.

The large fan turns **45 times each minute** for **80 minutes.**

$45 \times 80 = 3600$

Step 5: State the answer for part b.

The large fan makes **3600 rotations**.

Quick Check

The small fan spins **twice as fast** as the large fan, so it should make **twice as many turns** in the same time.

$28 \times 2 = 56$

That matches part **a)**.

For part **b)**, $45 \times 8 = 360$, so $45 \times 80 = 3600$.

This confirms the answer is correct.

Final Answer

The small fan turns **56 times**, and the large fan makes **3600 rotations** altogether.

📌 Common Mistake

*A common mistake is forgetting that part **a)** is about the fans turning for the **same time**, not adding or subtracting the two speeds.*

💡 Exam Tip

*When a question gives **rotations per minute**, think of it like a rate.*

*To find total rotations, multiply: **rotations per minute × number of minutes***

Combined Rates Comparison (rowing and swimming)

QUESTION

Mixi **rows** a boat to a small island and **swims back**. The total travel time is **27 minutes**.

If Mixi swam **both ways**, the total travel time would be **45 minutes**.

a) How long would the trip take if Mixi rowed both ways?

b) How many times as fast rowing is compared with swimming?

LET'S UNDERSTAND THE QUESTION TOGETHER

Mixi travels the **same distance** each way.

One trip is by **rowing**.
The other trip is by **swimming**.

We know the total time for the mixed trip is **27 minutes**.

We also know that if Mixi **swam both ways**, the total would be **45 minutes**.

So, we need to compare:

- ✓ A trip with **swimming both ways**.
- ✓ A trip with **rowing one way and swimming one way**.

Then we can work out the rowing time and compare the speeds.

WHAT WE KNOW

- ✓ Mixi rows to the island and swims back.
- ✓ The total time for that trip is **27 minutes**.
- ✓ If Mixi swam both ways, the total would be **45 minutes.**
- ✓ The distance each way is the same.

WHAT WE NEED TO FIND

We need to find:

1. How long the trip would take if Mixi **rowed both ways.**
2. How many times as fast **rowing** is compared with **swimming**.

Our Plan

1. First, find the time for **one swimming trip**.
2. Then use the mixed trip total to find the time for **one rowing trip**.
3. After that, work out:
   - The total time for rowing both ways.
   - The speed comparison between rowing and swimming.

Step-by-Step Solution

Step 1: Find the swimming time for one trip.

If Mixi swam both ways, the total time would be **45 minutes.**

So, one swimming trip takes: $45 \div 2 = 22.5$ **minutes.**

One swimming trip takes **22.5 minutes.**

Step 2: Use the mixed trip total.

We know:

rowing one way + swimming one way = 27 minutes

Swimming one way is **22.5 minutes**.

So: rowing one way + 22.5 = 27

Step 3: Find the rowing time for one trip.

rowing one way $= 27 - 22.5$ **minutes.**

rowing one way $= 4.5$

One rowing trip takes **4.5 minutes**.

Step 4: Find the time for rowing both ways.

One rowing trip is **4.5 minutes**.

So, rowing both ways would take:

$4.5 + 4.5 = 9$ **minutes.**

The trip would take **9 minutes** if Mixi rowed both ways.

Step 5: Compare rowing and swimming.

One swimming trip $= 22.5$ minutes

One rowing trip $= 4.5$ minutes

Both trips cover the same distance. **When the distance is the same, the method that takes less time is faster**.

Now compare the times:

$$22.5 \div 4.5 = 5$$

This means swimming takes 5 times as long as rowing for the same distance.
So, rowing is 5 times as fast as swimming.

Quick Check

Let's test the answers:

- ✓ Swimming both ways $= 22.5 + 22.5 =$ **45 minutes**
- ✓ Rowing one way and swimming one way $= 4.5 + 22.5 =$ **27 minutes**

Both totals match the question.

Final Answer

If Mixi rowed both ways, the trip would take **9 minutes**, and rowing is **5 times as fast** as swimming.

Common Mistake

*A common mistake is to divide **27 minutes** by 2 and assume each one-way trip takes the same time. That is not correct because **rowing and swimming are different speeds**.*

Exam Tip

*When one trip uses **two different speeds**, look for the information about **both ways at one speed** first. That helps you find the one-way time and makes the rest much easier.*

Data & Probability

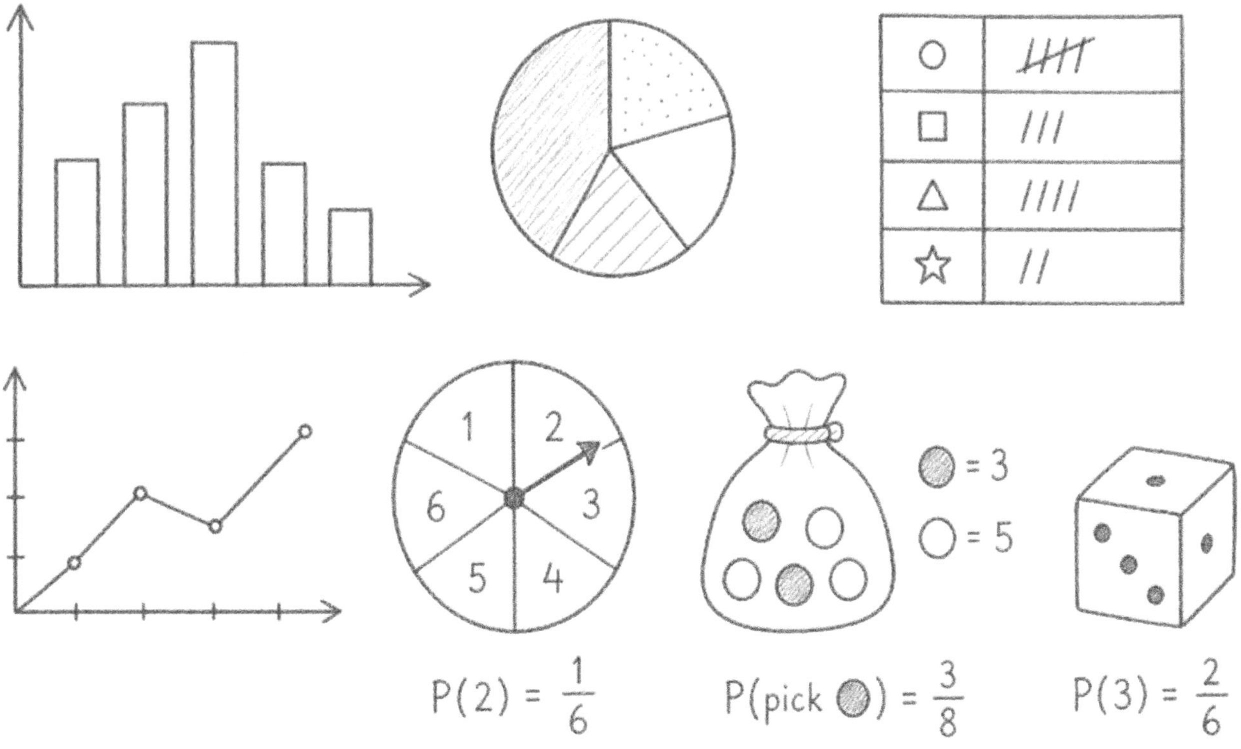

Probability with Part-Whole Counts (stickers)

QUESTION

Mika has star-shaped, circle-shaped, and square-shaped stickers.

There are **24 star-shaped stickers**. If the **probability** of picking a circle-shaped sticker is $\frac{1}{4}$, how many **square-shaped stickers** are there?

LET'S UNDERSTAND THE QUESTION TOGETHER

There are **60 stickers altogether**.

These stickers are split into **3 groups**: **star-shaped**, **circle-shaped**, and **square-shaped**.

We already know that **24 stickers are star-shaped**.

We are also told that the chance of picking a **circle-shaped** sticker is $\frac{1}{4}$. This means **one-quarter** of the 60 stickers are circles.

After we find the number of **circle-shaped** stickers, we can work out how many stickers are already counted.

Then the stickers that are **left** must be the **square-shaped** ones.

WHAT WE KNOW

1. Total number of stickers = **60**
2. Number of star-shaped stickers = **24**
3. Probability of picking a circle-shaped sticker = $\frac{1}{4}$

WHAT WE NEED TO FIND

We need to find the number of **square-shaped stickers.**

Our Plan

1. First, use the probability to find the number of **circle-shaped** stickers.
2. Next, add the **star-shaped** and **circle-shaped** stickers.
3. Then subtract that total from **60** to find the number of **square-shaped** stickers.

Step-by-Step Solution

Step 1: Find the number of circle-shaped stickers.

The probability of picking a circle-shaped sticker is:

$$\frac{\textbf{number of circle-shaped stickers}}{\textbf{total number of stickers}} = \frac{\mathbf{1}}{\mathbf{4}}$$

Since there are **60** stickers in total:

$$\text{number of circle-shaped stickers} = \frac{1}{4} \times 60 = 15$$

So, there are **15 circle-shaped stickers.**

Step 2: Add the star-shaped and circle-shaped stickers.

$24 + 15 = 39$

So, **39 stickers** are either star-shaped or circle-shaped.

Step 3: Subtract from the total to find the square-shaped stickers.

$60 - 39 = 21$

So, there are **21 square-shaped stickers.**

Quick Check

Let's check the probability:

$$\frac{15}{60} = \frac{1}{4}$$

That matches the question.

Now check the total:

$$24 + 15 + 21 = 60$$

Everything adds up correctly.

Final Answer

There are **21 square-shaped stickers.**

Common Mistake

*A common mistake is to subtract **24** from **60** first and forget to find the number of **circle-shaped** stickers from the probability.*

Exam Tip

*When probability is given as a fraction, think of it as **part over whole**. Use the total first, then find what is left at the end.*

Weighted Total from Frequency Table

QUESTION

The data below shows how many **fish tanks** have different numbers of fish.

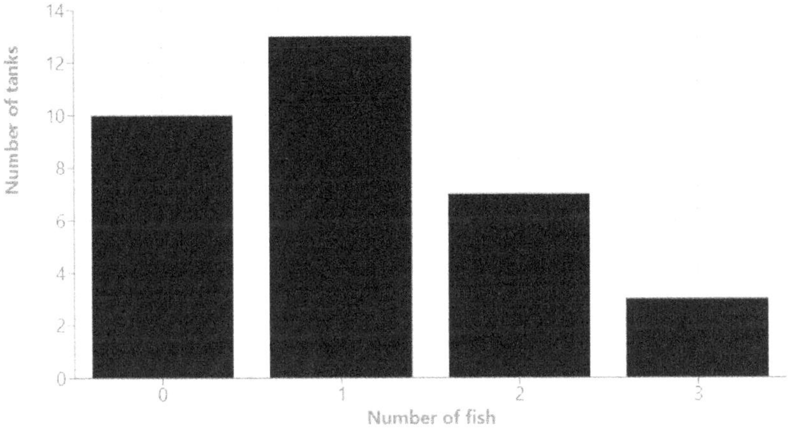

How many fish are there altogether in all the tanks?

LET'S UNDERSTAND THE QUESTION TOGETHER

We are given **groups of fish tanks**, and each group has a certain number of fish.

Some tanks have **no fish**, some have **1 fish**, some have **2 fish**, and some have **3 fish**.

Each group tells us:

- ✓ How many tanks there are.
- ✓ How many fish are in each tank.

We need to find the **total number of fish** in all the tanks combined.

WHAT WE KNOW

- ✓ 10 tanks have 0 fish.
- ✓ 13 tanks have 1 fish each.
- ✓ 7 tanks have 2 fish each.
- ✓ 3 tanks have 3 fish each.

WHAT WE NEED TO FIND

The total number of fish in all the tanks.

Our Plan

Multiply the number of tanks by the number of fish in each group, then add all the totals.

Step-by-Step Solution

Step 1: Fish in tanks with 0 fish

$10 \times 0 = \mathbf{0\ fish}$

Step 2: Fish in tanks with 1 fish

$13 \times 1 = \mathbf{13\ fish}$

Step 3: Fish in tanks with 2 fish

$7 \times 2 = \mathbf{14\ fish}$

Step 4: Fish in tanks with 3 fish

$3 \times 3 = \mathbf{9\ fish}$

Step 5: Add all the fish together

$0 + 13 + 14 + 9 = \mathbf{36}$

Quick Check

13 tanks already give 13 fish.

Adding the other groups gives a total of **36**. We counted fish, not tanks, because each bar shows how many tanks are in each group.

Final Answer

There are **36 fish altogether** in all the tanks.

Common Mistake

*Some students add **10 + 13 + 7 + 3** and get the number of tanks instead of fish. Remember to **multiply first**, then add.*

Exam Tip

When you see grouped data like this, always think:
***number of groups × amount in each group**, then add.*

Average with a Missing Value

QUESTION

Mika has these number cards.

She removes one card. The average of the numbers on the remaining cards is 11.

Which number card did Mika remove?

LET'S UNDERSTAND THE QUESTION TOGETHER

Mika starts with **7 number cards**.

Then she takes away **1 card**, so **6 cards are left**.

We are told that the **average** of the 6 remaining numbers is **11**.

Our job is to find the **missing card**.

WHAT WE KNOW

- ✓ The number cards are 4, 6, 8, 10, 12, 14, 16
- ✓ There are 7 cards at the start.
- ✓ 1 card is removed.
- ✓ 6 cards remain.
- ✓ The average of the remaining 6 cards is 11

WHAT WE NEED TO FIND

We need to find **which number card was removed**.

Our Plan

1. First, find the total of all the cards.
2. Next, find the total of the 6 cards left by using the average.
3. Then subtract to work out the missing card.

Step-by-Step Solution

Step 1: Find the total of all 7 number cards.

$4 + 6 + 8 + 10 + 12 + 14 + 16 = 70$

So, the total at the start is **70**.

Step 2: Find the total of the 6 cards left.

Average = total ÷ number of cards

So:

Total = average × number of cards

Total $= 11 \times 6 = 66$

So, the total of the remaining cards is **66**.

Step 3: Find the removed card.

$70 - 66 = 4$

So, the removed card is **4**.

Quick Check

If Mika removes **4**, the remaining cards are:

6, 8, 10, 12, 14, 16

Their total is: $6 + 8 + 10 + 12 + 14 + 16 =$ **66**

Now check the average: $66 \div 6 =$ **11**

It works.

Final Answer

Mika removed the card with **4** on it.

Common Mistake

*A common mistake is using **7 cards** when finding the new total.*
*But after one card is removed, there are only **6 cards left**.*

Exam Tip

*When an average question has a **missing number**, try working with the **total** first.*
That usually makes the missing value much easier to find.

Probability from Ratio Relationships (stickers)

QUESTION

There are the same number of flower stickers as star stickers. There are 3 times as many star stickers as smiley-face stickers. One sticker is picked from the box without looking.

What fraction of the stickers are star stickers?

LET'S UNDERSTAND THE QUESTION TOGETHER

We have **3 types of stickers**:

- ✓ Flower stickers.
- ✓ Star stickers.
- ✓ Smiley-face stickers.

The question tells us how the numbers are related.

The number of **flower stickers** is the **same** as the number of **star stickers**.

The number of **star stickers** is **3 times** the number of **smiley-face stickers**.

We are not told the exact number of stickers, but we can still work out the probability because we know the **ratio** between the groups.

We need to find the chance that the sticker picked is a **star sticker**.

WHAT WE KNOW

- ✓ Flower stickers = Star stickers.
- ✓ Star stickers = 3 × Smiley-face stickers.
- ✓ Probability = number of star stickers ÷ total number of stickers.

WHAT WE NEED TO FIND

We need to find the **probability of picking a star sticker**.

Our Plan

We will:

1. Treat the number of smiley-face stickers as 1 part.
2. Use the clues to work out the number of star and flower stickers.
3. Find the total number of stickers.
4. Write the probability as

$$\frac{\text{number of star stickers}}{\text{total number of stickers}}$$

Step-by-Step Solution

Step 1: Let the number of smiley-face stickers be 1 part

Since the number of star stickers is **3 times** the number of smiley-face stickers, we can use a simple ratio.

So:

- ✓ Smiley-face stickers = 1 part
- ✓ Star stickers = 3 parts

Step 2: Work out the number of flower stickers

The number of flower stickers is the **same** as the number of star stickers.

So: Flower stickers = 3 parts

Step 3: Add all the parts

Now we have:

- ✓ Flower = 3 parts
- ✓ Star = 3 parts
- ✓ Smiley-face = 1 part

Total parts: $3 + 3 + 1 = 7$
So, there are **7 equal parts** altogether.

Step 4: Find the probability of picking a star sticker

The number of star-sticker parts is 3.

So, the probability is: $\frac{3}{7}$

Quick Check

Let's test it with actual numbers.

If there are:

- ✓ 1 smiley-face sticker
- ✓ 3 star stickers
- ✓ 3 flower stickers

Then total stickers: $1 + 3 + 3 = 7$

Probability of picking a star sticker: $\frac{3}{7}$

Final Answer

The probability of picking a star sticker is $\mathbf{\frac{3}{7}}$.

Common Mistake

*A common mistake is to think the answer is $\frac{1}{3}$ because there are 3 types of stickers. That is not correct because the 3 types do **not** have the same number of stickers.*

Exam Tip

*When a question gives relationships like **"same number as"** or **"3 times as many"**, think in **parts or ratios** first. You do not need exact numbers to find the probability.*

Conditional Probability After Removal (fruits)

QUESTION

Mika randomly picks **2 fruits** from the basket.

One is an **apple** and one is an **orange**, and she **keeps them out**. She then picks **1 more fruit**. What is the **probability** that it is an **orange**?

LET'S UNDERSTAND THE QUESTION TOGETHER

Mika starts with **two kinds of fruit** in the basket.

She takes out **2 fruits first**, and we are told exactly what they are:

- ✓ **1 apple**
- ✓ **1 orange**

So, the number of apples goes down by **1**, and the number of oranges also goes down by **1**.

Then Mika picks **one more fruit**.

We need to find the chance that this next fruit is **an orange**.

WHAT WE KNOW

- ✓ There are **10 apples** at the start.
- ✓ There are **6 oranges** at the start.
- ✓ Mika removes:
  - **1 apple**
  - **1 orange**
- ✓ The fruits are **not put back**.

WHAT WE NEED TO FIND

The probability that the **next fruit picked is an orange**.

Our Plan

First, work out how many apples and oranges are left.

Then work out how many fruits are left altogether.

Finally, write the probability as:

$$\text{Probability} = \frac{\text{number of oranges left}}{\text{total fruits left}}$$

Step-by-Step Solution

Step 1: Work out how many apples are left.

Mika started with 10 apples.

She took out 1 apple.

$10 - 1 = 9$

So, **9 apples** are left.

Step 2: Work out how many oranges are left.

Mika started with 6 oranges.

She took out 1 orange.

$6 - 1 = 5$

So, **5 oranges** are left.

Step 3: Work out how many fruits are left altogether.

$9 + 5 = 14$

So, there are **14 fruits** left in the basket.

Step 4: Find the probability of picking an orange next.

There are 5 oranges left out of 14 fruits.

$$\text{Probability} = \frac{5}{14}$$

Quick Check

We can check by using the **numbers left in the basket.**

After 1 apple and 1 orange are removed, there are:

✓ **9 apples**

✓ **5 oranges**

That means there are **14 fruits** left altogether.

Since **5 out of the 14** remaining fruits are oranges, the probability of picking an orange next is: $\mathbf{\frac{5}{14}}$

So, our answer is correct.

Final Answer

The probability that the next fruit is an orange is $\frac{5}{14}$.

Common Mistake

A common mistake is forgetting to ***change both groups*** *after the first two fruits are taken out. You must remove* ***1 apple*** *and* ***1 orange*** *before finding the new probability.*

Exam Tip

When the question tells you exactly what was taken out, always ***update the numbers first*** *before you calculate the probability.*

Probability with Decimal Comparison

QUESTION

Mika has a bag of 10 cards (shown in the image). Each card has a decimal written on it.

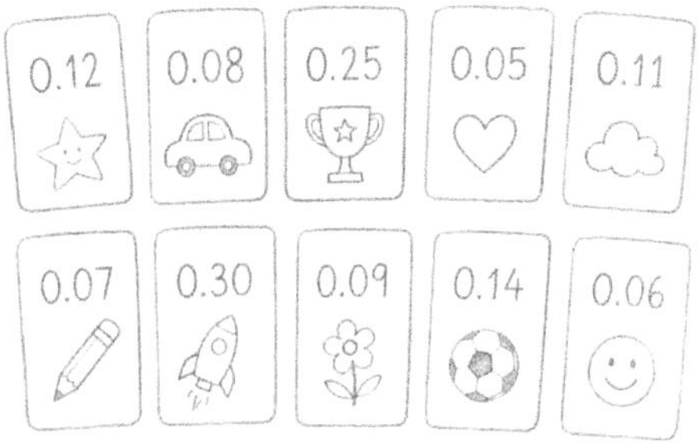

Mika picks one card at random. What is the probability that the number on the card is greater than $\frac{1}{10}$?

LET'S UNDERSTAND THE QUESTION TOGETHER

Each card has an **equal chance** of being chosen.
We are looking for numbers that are **greater than** $\frac{1}{10}$.

$$\frac{1}{10} = 0.1$$

So, we are really finding how many numbers are **greater than 0.1**.

WHAT WE KNOW

- ✓ There are **10 cards in total**.
- ✓ Each card is equally likely.
- ✓ We are comparing numbers to **0.1**

WHAT WE NEED TO FIND

The **probability** that a randomly chosen card has a number **greater than 0.1**

Our Plan

1. Convert the fraction to a decimal.
2. Identify all numbers greater than that value.
3. Count how many matches.
4. Write probability as a fraction.

Step-by-Step Solution

Step 1: Convert the fraction

$$\frac{1}{10} = 0.1$$

Step 2: List numbers greater than 0.1
From the list:

- ✓ 0.12
- ✓ 0.25
- ✓ 0.11
- ✓ 0.30
- ✓ 0.14

Step 3: Count them

There are **5 numbers** greater than 0.1

Step 4: Count total cards

There are **10 cards**

Step 5: Write the probability

$$\text{Probability} = \frac{5}{10}$$

Step 6: Simplify

$$\frac{5}{10} = \frac{1}{2}$$

Quick Check

Half the numbers are above 0.1, so a probability of $\frac{1}{2}$ makes sense.

Final Answer

The probability is $\frac{1}{2}$.

📌 Common Mistake

*Including **0.1 itself** as a valid number.*
*Be careful: the question says **greater than**, not equal to!*

💡 Exam Tip

*Always **convert fractions to decimals (or vice versa)** before comparing. It makes spotting the correct values much easier.*

Finding a Missing Value (Line Graph)

QUESTION

A line chart shows the number of students in **Groups A to F** at a sports centre. The point for **Group F has not been shown**.

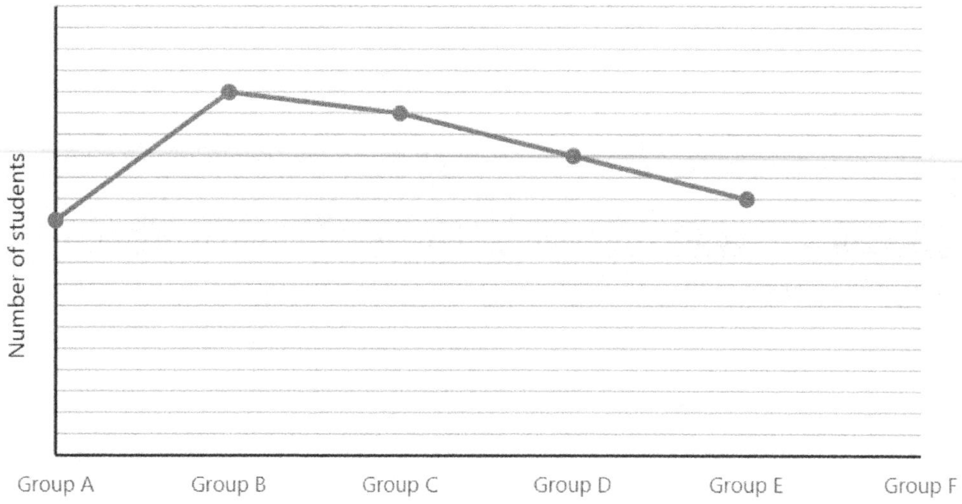

The total number of students across all six groups is **180**.

How many students are in Group F?

LET'S UNDERSTAND THE QUESTION TOGETHER

There are **6 groups altogether**.
The line chart gives the numbers for **5 groups only**.
That means **1 group is missing**.

We also know the **total number of students is 180**.
So the missing number must be the amount needed to make the total **180**.

WHAT WE KNOW

Total students in all 6 groups = **180**

- ✓ Group A = **22**
- ✓ Group B = **34**
- ✓ Group C = **32**
- ✓ Group D = **28**
- ✓ Group E = **24**

WHAT WE NEED TO FIND

The number of students in **Group F.**

Our Plan

1. First, add the students in Groups A to E.
2. Then subtract that total from 180.
3. The answer will be the number of students in Group F.

Step-by-Step Solution

Step 1: Add Group A and Group B.

$22 + 34 = 56$

Step 2: Add Group C.

$56 + 32 = 88$

Step 3: Add Group D.

$88 + 28 = 116$

Step 4: Add Group E.

$116 + 24 = 140$

Step 5: Subtract from the total number of students.

$180 - 140 = 40$

So, **Group F has 40 students**.

Quick Check

Let's add all 6 groups to check:

$$22 + 34 + 32 + 28 + 24 + 40 = 180$$

The total is correct, so our answer makes sense.

Final Answer

There are **40 students in Group F**.

Common Mistake

A common mistake is to forget one of the groups when adding the known numbers. That gives the wrong leftover amount.

Exam Tip

*When one part is missing and the **whole total** is given, add the known parts first, then subtract from the total.*

Finding a Missing Frequency (Dot Plot)

QUESTION

A dot plot shows how many books 36 students read last month.

From the dot plot, Nai records this information:

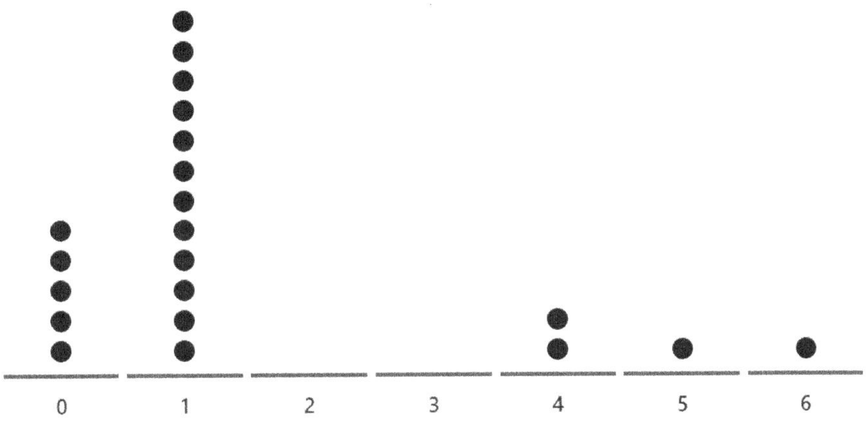

The number of students who read **2 books** is **twice** the number of students who read **3 books**.

How many students read **2 books**?

LET'S UNDERSTAND THE QUESTION TOGETHER

We know the total number of students is **36**.

Some of the groups are already given.

The missing groups are the students who read **2 books** and **3 books**.

We are told that the number of students who read **2 books** is **twice** the number of students who read **3 books**.

So, we need to use the total number of students and the missing relationship to find the 2-book group.

WHAT WE KNOW

- ✓ Total students = **36**
- ✓ 5 students read **0 books**.
- ✓ 12 students read **1 book.**
- ✓ 2 students read **4 books**.
- ✓ 1 student read **5 books**.
- ✓ 1 student read **6 books**.
- ✓ The number of students who read **2 books** is **twice** the number who read **3 books.**

WHAT WE NEED TO FIND

How many students read **2 books.**

Our Plan

1. First, add the groups we already know.
2. Then subtract from the total of 36 to find how many students are left.
3. Finally, split the remaining students using the ratio **2:1**.

Step-by-Step Solution

Step 1: Add the known groups

$5 + 12 + 2 + 1 + 1 = 21$

So, **21 students** are already counted.

Step 2: Find how many students are left

$36 - 21 = 15$

So, **15 students** read either **2 books** or **3 books**.

Step 3: Write the relationship as a ratio

The number who read **2 books** is twice the number who read **3 books**.

So, the ratio is: 2: 1

That means the 15 students are split into **3 equal parts.**

Step 4: Find the value of 1 part

$15 \div 3 = 5$

So, **1 part = 5 students**.

Step 5: Find the number who read 2 books

The 2-book group is **2 parts**.

$2 \times 5 = 10$

So, **10 students** read 2 books.

Quick Check

If the 3-book group is **5**, then the 2-book group is **10**.

Now check the total: $5 + 12 + 10 + 5 + 2 + 1 + 1 = 36$

The total is correct.

Final Answer

10 students read 2 books.

Common Mistake

A common mistake is to forget to subtract the known groups first.
*You must find how many students are left for the **2-book** and **3-book** groups before using the ratio.*

Exam Tip

*When one group is **twice** another group, think of a **2 : 1 ratio** straight away. Then divide the leftover total into equal parts.*

Ordering Fractions by Size

QUESTION

At a school fruit stall, five students each ate part of a watermelon. The amounts they ate are shown below:

- ✓ Mika: $\frac{3}{10}$
- ✓ Nai: $\frac{7}{20}$
- ✓ Ava: $\frac{2}{5}$
- ✓ Ben: $\frac{1}{4}$
- ✓ Zara: $\frac{9}{20}$

Which three students ate the **smallest amounts**? Write their names in order, starting with the smallest amount.

LET'S UNDERSTAND THE QUESTION TOGETHER

Each student ate a **fraction** of the same watermelon.

- ✓ The fractions show **how much** each student ate.
- ✓ We need to compare the fractions carefully.
- ✓ We are looking for the **three smallest amounts**.
- ✓ Then we write the names in order from the **smallest fraction** to the next smallest.

WHAT WE KNOW

- ✓ Mika: $\frac{3}{10}$
- ✓ Nai: $\frac{7}{20}$
- ✓ Ava: $\frac{2}{5}$
- ✓ Ben: $\frac{1}{4}$
- ✓ Zara: $\frac{9}{20}$

WHAT WE NEED TO FIND

- ✓ The **three smallest fractions**.
- ✓ The students' names in the correct order.

Our Plan

1. Change the fractions so they have the **same denominator**.
2. Compare the numerators.
3. Put the fractions in order from smallest to largest.
4. Write the first three names.

Step-by-Step Solution

Step 1: Choose a common denominator

The denominators are 10, 20, 5, 4 and 20.

A common denominator is **20**.

Step 2: Rewrite each fraction with denominator 20

- ✓ Mika: $\frac{3}{10} = \frac{6}{20}$
- ✓ Nai: $\frac{7}{20} = \frac{7}{20}$
- ✓ Ava: $\frac{2}{5} = \frac{8}{20}$
- ✓ Ben: $\frac{1}{4} = \frac{5}{20}$
- ✓ Zara: $\frac{9}{20} = \frac{9}{20}$

Step 3: Compare the fractions

From smallest to largest, they are:

1. $\frac{5}{20}$
2. $\frac{6}{20}$

3. $\frac{7}{20}$

4. $\frac{8}{20}$

5. $\frac{9}{20}$

Step 4: Match the fractions to the students

✓ $\frac{5}{20} = \frac{1}{4} \rightarrow$ Ben

✓ $\frac{6}{20} = \frac{3}{10} \rightarrow$ Mika

✓ $\frac{7}{20} \rightarrow$ Nai

Quick Check

Let's check with decimals:

✓ $\frac{1}{4} = 0.25$

✓ $\frac{3}{10} = 0.3$

✓ $\frac{7}{20} = 0.35$

These do go from smallest to larger, so our order is correct.

Final Answer

The three students who ate the smallest amounts are **Ben, Mika and Nai**, in that order.

Common Mistake

A common mistake is comparing only the ***bottom numbers*** *of the fractions.*
You cannot do that unless the fractions already have the ***same denominator****.*

Exam Tip

When comparing fractions, first turn them into ***equivalent fractions with the same denominator****. Then it becomes much easier to put them in order.*

Time-Series Dot Plot Analysis (ferries)

QUESTION

A **dot plot** shows the number of ferries at the terminal at each hour from **8:00 am** to **4:00 pm**.

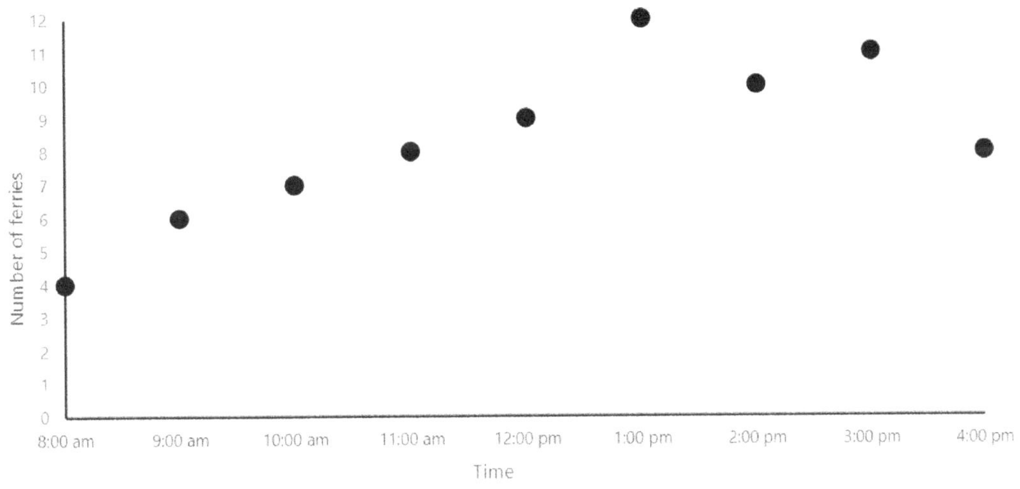

Nai started work at **8:00 am.**

How long after Nai started work was the number of ferries at the terminal **triple** the number at **8:00 am**?

LET'S UNDERSTAND THE QUESTION TOGETHER

We are told how many ferries are at the terminal at different times.

At **8:00 am**, there are **4 ferries**.

The question asks when the number becomes **triple** that starting number.

So first we need to think about what **triple 4** means.

Then we need to find the time when the dot plot shows that number.

After that, we work out **how much time passed after 8:00 am**.

WHAT WE KNOW

- ✓ Nai started work at **8:00 am.**
- ✓ At **8:00 am**, there are **4 ferries.**
- ✓ We need to find when the number of ferries is **triple** that amount.
- ✓ Triple means **3 times as many.**

WHAT WE NEED TO FIND

We need to find **how long after 8:00 am** the number of ferries was **triple 4.**

Our Plan

First, work out what **triple 4** is.

Next, find the time when the dot plot shows that number.

Then, calculate the time from **8:00 am** to that time.

Step-by-Step Solution

Step 1: Find the number of ferries at 8:00 am.

At **8:00 am**, there are **4 ferries.**

Step 2: Work out triple that number.

Triple means multiply by 3:

$4 \times 3 = 12$

So, we are looking for the time when there are **12 ferries.**

Step 3: Find that number in the dot plot values.

The dot plot shows **12 ferries at 1:00 pm.**

Step 4: Work out how long it is from 8:00 am to 1:00 pm.

From **8:00 am to 1:00 pm** is **5 hours.**

Quick Check

At **8:00 am**, there are **4 ferries**.

Triple **4** is **12**.

The dot plot shows **12 ferries at 1:00 pm**.

From **8:00 am to 1:00 pm** is **5 hours**, so the answer fits.

Final Answer

The number of ferries was triple the 8:00 am amount **5 hours after Nai started work.**

Common Mistake

A common mistake is giving the answer as ***1:00 pm****.*
But the question asks ***how long after*** *Nai started work, so the answer should be* ***5 hours****.*

Exam Tip

When you see words like ***double*** *or* ***triple****, calculate that target number first. Then find the matching time, and only after that work out the* ***elapsed time****.*

Categorical Data Analysis (school activities)

QUESTION

A teacher asked students in **Nai's year group** which after-school activity they do on Thursdays. The results are shown below:

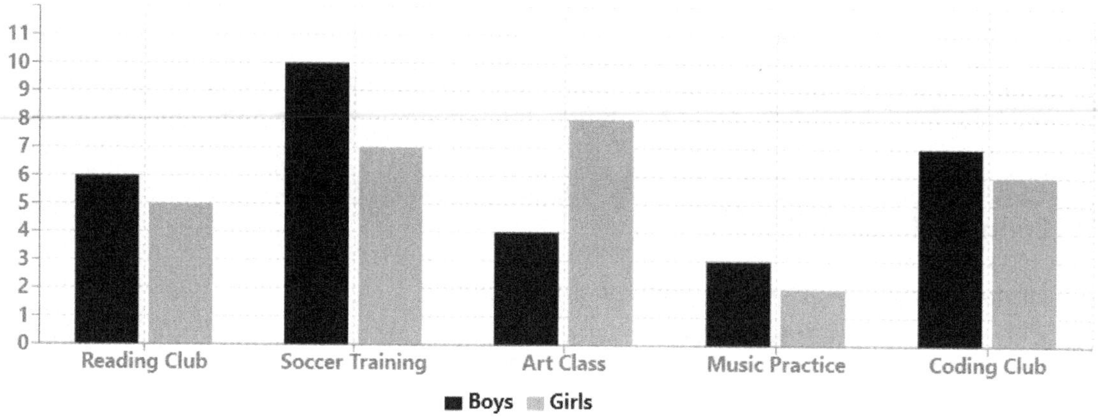

a) How many students are in the year group altogether?

b) How many more boys than girls go to either **soccer training, music practice, or coding club**?

c) How many students go to **art class or reading club** altogether?

d) True or false: **More students go to soccer training than to art class and music practice combined.**

LET'S UNDERSTAND THE QUESTION TOGETHER

This question gives us the numbers of **boys and girls** in five different after-school activities.

Each part asks us to do something a little different.

- ✓ In **a)**, we need the total number of students
- ✓ In **b)**, we only use **soccer training, music practice, and coding club**
- ✓ In **c)**, we only use **art class and reading club**
- ✓ In **d)**, we compare **soccer training** with **art class and music practice together**

We need to read each part carefully and only use the activities named in that part.

WHAT WE KNOW

- ✓ Reading club: 6 boys, 5 girls.
- ✓ Soccer training: 10 boys, 7 girls.
- ✓ Art class: 4 boys, 8 girls.
- ✓ Music practice: 3 boys, 2 girls.
- ✓ Coding club: 7 boys, 6 girls.

WHAT WE NEED TO FIND

We need to find:

- ✓ The total number of students.
- ✓ The difference between boys and girls in soccer training, music practice, and coding club.
- ✓ The total number of students in art class and reading club.
- ✓ Whether the statement in part d) is true or false.

Our Plan

1. Find the total for each activity.
2. Add the correct groups for each part.
3. Compare totals carefully in the last part.

Step-by-Step Solution

Step 1: Find the total number of students in each activity.

Reading club: $6 + 5 = 11$
Soccer training: $10 + 7 = 17$
Art class: $4 + 8 = 12$
Music practice: $3 + 2 = 5$
Coding club: $7 + 6 = 13$

Step 2: Find the total number of students altogether.

$11 + 17 + 12 + 5 + 13 = 58$

So, there are **58 students** in the year group.

Step 3: Add the boys in soccer training, music practice, and coding club.

$10 + 3 + 7 = 20$ boys

Step 4: Add the girls in soccer training, music practice, and coding club.

$7 + 2 + 6 = 15$ girls

Step 5: Find how many more boys there are.

$20 - 15 = 5$

So, there are **5 more boys.**

Step 6: Find the total number of students in art class and reading club.

$12 + 11 = 23$

So, **23 students** go to art class or reading club altogether.

Step 7: Compare soccer training with art class and music practice combined.

Soccer training = 17

Art class and music practice = $12 + 5 = 17$

The two totals are equal, so the statement is **false.**

Quick Check

Let's check the last part again:

✓ Soccer training = 17

✓ Art class + music practice = 17

They are the same, so it is **not more.** That confirms the statement is false.

Final Answer

a) There are **58 students** altogether.

b) There are **5 more boys than girls.**

c) **23 students** go to art class or reading club altogether.

d) The statement is **false.**

📌 Common Mistake

A common mistake is using the wrong activities for each part. Always check exactly which groups the question tells you to use.

Exam Tip

In multi-part questions, solve ***one part at a time*** *and tick off the activity names as you use them. This helps stop mix-ups.*

Cumulative Totals Comparison (stamps)

QUESTION

Mika is collecting stamps over 5 weeks for a school project. Mika starts with **0 stamps**.

At the end of each week, the **total number of stamps** is recorded:

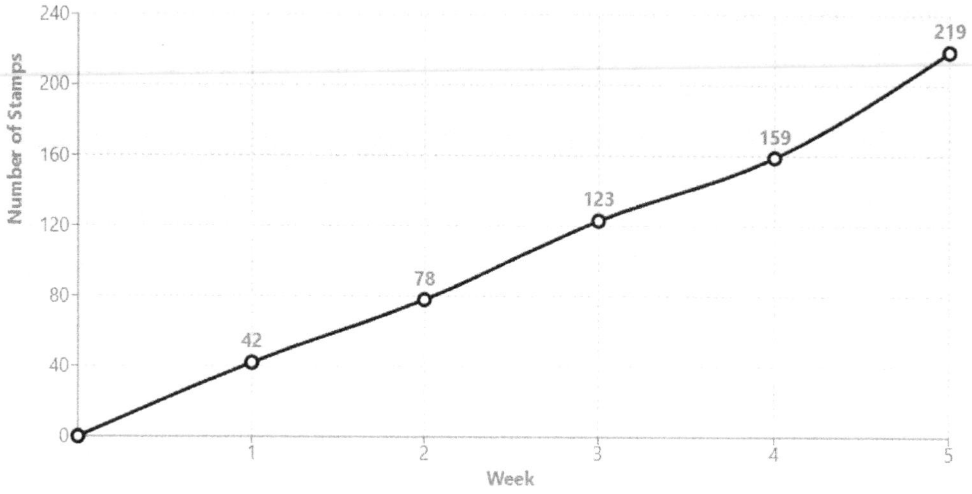

a) During which week did Mika collect the **most stamps**?

b) During which **two weeks** did Mika collect **exactly the same number of stamps**?

LET'S UNDERSTAND THE QUESTION TOGETHER

These numbers are **totals** at the end of each week.

That means the number for week 3 includes stamps from weeks 1, 2 and 3.

So, we are not looking for the biggest total.

We are looking for how many stamps were added **in each single week.**

Then we need to answer **two questions**:

- ✓ Which week had the **greatest increase**.
- ✓ Which **two weeks** had the **same increase.**

WHAT WE KNOW

- ✓ Mika starts with **0 stamps**
- ✓ The totals at the end of each week are:
  - Week 1: **42**
  - Week 2: **78**
  - Week 3: **123**
  - Week 4: **159**
  - Week 5: **219**

WHAT WE NEED TO FIND

- ✓ How many stamps Mika collected in **each week**.
- ✓ Which week had the **most stamps collected**.
- ✓ Which **two weeks** had **equal amounts collected.**

Our Plan

1. We will subtract each total from the one before it.
2. That will show how many stamps were collected in each week.
3. Then we will compare the weekly amounts.

Step-by-Step Solution

Step 1: Find the number of stamps collected in week 1.

Mika started with 0 stamps.

So: $42 - 0 = \mathbf{42}$

Mika collected **42 stamps** in week 1.

Step 2: Find the number of stamps collected in week 2.

$78 - 42 = \mathbf{36}$

Mika collected **36 stamps** in week 2.

Step 3: Find the number of stamps collected in week 3.

$123 - 78 = \mathbf{45}$

Mika collected **45 stamps** in week 3.

Step 4: Find the number of stamps collected in week 4.

$159 - 123 = \mathbf{36}$

Mika collected **36 stamps** in week 4.

Step 5: Find the number of stamps collected in week 5.

$219 - 159 = \mathbf{60}$

Mika collected **60 stamps** in week 5.

Step 6: List the weekly amounts clearly.

- ✓ Week 1: 42 stamps
- ✓ Week 2: 36 stamps
- ✓ Week 3: 45 stamps
- ✓ Week 4: 36 stamps
- ✓ Week 5: 60 stamps

Step 7: Find the week with the greatest amount collected.

The largest weekly amount is **60 stamps.**

So, Mika collected the most stamps in **week 5**.

Step 8: Find the two weeks with exactly the same amount collected.

Week 2 = **36 stamps**
Week 4 = **36 stamps**

So, the two weeks with the same amount are **week 2 and week 4**.

Quick Check

Add the weekly amounts:

$42 + 36 + 45 + 36 + 60 = \mathbf{219}$

This matches the final total, so the working is correct.

Final Answer

Mika collected the most stamps in **week 5**, and the two weeks with exactly the same amount collected were **week 2 and week 4**.

📌 Common Mistake

*A common mistake is comparing the **totals** instead of the **weekly increases**.*
*Another common mistake is forgetting that the first week must be compared with **0** because Mika started with none.*

💡 Exam Tip

*When you see totals recorded over time, first work out the **change between each pair of weeks**.*
*After that, questions about the **most**, **least**, or **same amount** become much easier.*

A Final Word Before You Step Forward

First, well done.

You should be proud of how far you've come.

Completing this book is a great achievement. It takes effort, focus, and the willingness to keep working even when things feel challenging. At this stage, your preparation is not just about practising questions. It is about trusting your thinking and staying steady under pressure.

There will be questions in the exam that feel familiar, and some that may feel new. When that happens, remember what you have practised. Pause, think, and break the problem down. You already know how to do this.

Not every question needs to be perfect. What matters is how you approach each one. Take a moment to recognise how far you have come, and the support you have had along the way.

A problem can often be solved in more than one way. You may find your own method, and that's a great thing. Like crossing a river, there are many paths you can take. What matters is reaching the other side. The result matters, but so does your thinking. Find your feet, trust your approach, and keep going.

You have done the work. Stay calm, trust yourself, and go and show what you can do.

Good luck.

Made in the USA
Monee, IL
05 November 2026

67820202R00269